SpringerBriefs in Applied Sciences and Technology

SpringerBriefs present concise summaries of cutting-edge research and practical applications across a wide spectrum of fields. Featuring compact volumes of 50 to 125 pages, the series covers a range of content from professional to academic.

Typical publications can be:

- A timely report of state-of-the art methods
- An introduction to or a manual for the application of mathematical or computer techniques
- A bridge between new research results, as published in journal articles
- A snapshot of a hot or emerging topic
- An in-depth case study
- A presentation of core concepts that students must understand in order to make independent contributions

SpringerBriefs are characterized by fast, global electronic dissemination, standard publishing contracts, standardized manuscript preparation and formatting guidelines, and expedited production schedules.

On the one hand, **SpringerBriefs in Applied Sciences and Technology** are devoted to the publication of fundamentals and applications within the different classical engineering disciplines as well as in interdisciplinary fields that recently emerged between these areas. On the other hand, as the boundary separating fundamental research and applied technology is more and more dissolving, this series is particularly open to trans-disciplinary topics between fundamental science and engineering.

Indexed by EI-Compendex, SCOPUS and Springerlink.

Zulhazmee Bakri · Zakiah Ahmad

Charring Rates of Laminated Veneer Lumber (LVL) from Tropical Hardwood Timber

Zulhazmee Bakri
Faculty of Engineering, Science
and Technology (FEST)
Infrastructure University Kuala Lumpur
(IUKL)
Kajang, Selangor, Malaysia

Zakiah Ahmad
Faculty of Civil Engineering
Universiti Teknologi MARA (UiTM)
Shah Alam, Selangor, Malaysia

ISSN 2191-530X   ISSN 2191-5318 (electronic)
SpringerBriefs in Applied Sciences and Technology
ISBN 978-981-96-5836-7   ISBN 978-981-96-5837-4 (eBook)
https://doi.org/10.1007/978-981-96-5837-4

© The Editor(s) (if applicable) and The Author(s), under exclusive license to Springer Nature Singapore Pte Ltd. 2025

This work is subject to copyright. All rights are solely and exclusively licensed by the Publisher, whether the whole or part of the material is concerned, specifically the rights of translation, reprinting, reuse of illustrations, recitation, broadcasting, reproduction on microfilms or in any other physical way, and transmission or information storage and retrieval, electronic adaptation, computer software, or by similar or dissimilar methodology now known or hereafter developed.

The use of general descriptive names, registered names, trademarks, service marks, etc. in this publication does not imply, even in the absence of a specific statement, that such names are exempt from the relevant protective laws and regulations and therefore free for general use.

The publisher, the authors and the editors are safe to assume that the advice and information in this book are believed to be true and accurate at the date of publication. Neither the publisher nor the authors or the editors give a warranty, expressed or implied, with respect to the material contained herein or for any errors or omissions that may have been made. The publisher remains neutral with regard to jurisdictional claims in published maps and institutional affiliations.

This Springer imprint is published by the registered company Springer Nature Singapore Pte Ltd.
The registered company address is: 152 Beach Road, #21-01/04 Gateway East, Singapore 189721, Singapore

If disposing of this product, please recycle the paper.

Preface

Fire safety is a critical consideration in the design and construction of timber structures. As engineered wood products such as laminated veneer lumber (LVL) continue to gain popularity due to their strength, sustainability, and versatility, understanding their performance in fire conditions has become increasingly important. One key aspect of fire performance is the charring rate, which directly influences the load-bearing capacity and structural integrity of LVL during a fire event. The charring rate is the rate at which the timber's cross-section is reduced when exposed to fire.

Internationally, fire safety standards such as Eurocode 5 (EC5) provide standardized charring rates for different timber materials. For LVL with a density above 480 kg/m^3, EC5 specifies a one-dimensional charring rate of 0.65 mm/min and a two-dimensional or notional rate of 0.70 mm/min. While these values serve as important benchmarks, they may not accurately reflect the fire performance of all timber species, particularly tropical hardwoods, which can exhibit distinct density and combustion characteristics.

In Malaysia, the national standard MS 544 Part 9:2001 classifies charring rates based on the strength grade of solid timber. However, this classification lacks specific fire performance data for LVL, despite the material's increasing adoption in modern construction. Given the growing use of LVL as a sustainable alternative to solid timber, it is imperative to investigate its fire behavior, particularly for species commonly found in Malaysia's tropical forests.

This book aims to address this knowledge gap by systematically examining the charring characteristics of LVL manufactured from selected Malaysian hardwood species, including Kedondong, Eucalyptus, Mengkulang, Rubberwood, and Kasai. This work is the result of extensive research and experimentation through a series of controlled fire tests; both one-dimensional and two-dimensional charring behaviors are analyzed. Fire resistance tests are conducted in accordance with BS 476: Part 22, while charring rate determination follows prEN 13381-7:2014. The results provide valuable insights into the relationship between timber density and fire performance, demonstrating that denser species generally exhibit lower charring rates. Additionally, scale effects are identified as a significant factor influencing the overall fire resistance of LVL.

By providing a comprehensive understanding of the fire behavior of Malaysian LVL, this book will serve as a useful reference for anyone involved in the field of fire safety, structural engineering, policymakers, and sustainable timber construction. It contributes to the advancement of fire safety regulations, supports the development of localized fire design guidelines, and promotes the safe and effective application of LVL in timber construction. It is hoped that the findings presented in this book will facilitate the broader adoption of LVL in structural applications while ensuring fire safety remains a top priority in timber construction.

Kajang, Malaysia	Zulhazmee Bakri
Shah Alam, Malaysia	Zakiah Ahmad

Competing Interests The authors have no competing interests to declare that are relevant to the content of this manuscript.

Contents

1 Engineered Wood Products as Construction Material 1
   1.1 Introduction 1
   1.2 Benefit 3
   1.3 Laminated Veneer Lumber (LVL) 7
   1.4 Advantages of Laminated Veneer Lumber 11
   1.5 Laminated Veneer Lumber Design Based on Malaysian Standard ... 15
      1.5.1 Design Method 15
      1.5.2 Design Considerations 17
   1.6 Conclusions 17
   References 18

2 Fundamentals of Fire and Combustion 21
   2.1 Introduction 21
   2.2 Composition of Wood 23
   2.3 Fire Behavior of Wood: A Molecular Perspective 25
   2.4 The Chemistry of Combustion 27
   2.5 Phases of Combustion 27
   2.6 Factors Affecting Wood Combustion Efficiency 33
   2.7 Summary 34
   References 35

3 Fire Resistance 39
   3.1 Introduction 39
   3.2 Building Fire Requirements 40
   3.3 Malaysian Uniform Building by Law 41
   3.4 Fire Resistance 42
   3.5 Reaction to Fire 45
   3.6 Reaction to Fire Classification 48

	3.7	Charring of Wood	52
		3.7.1 One-Dimensional Charring Rate	54
		3.7.2 Two-Dimensional Charring Rate	55
	3.8	Fire Resistance Considerations in Timber Structure Design: The Malaysian Context	57
	3.9	Summary	60
	References		61
4	**One-Dimensional Charring Rate for Laminated Veneer Lumber from Malaysian Tropical Timber**		**63**
	4.1	Introduction	63
	4.2	Testing Material, Specimens, and Equipment	64
		4.2.1 Laminated Veneer Lumber	64
		4.2.2 Furnace Preparation	64
		4.2.3 Furnace Cover Preparation	65
		4.2.4 Preparation of Specimen and Data Acquisition Equipment	65
	4.3	Visual Observations	69
	4.4	Determining the Residual Cross-Section: Measurement and Analysis	69
	4.5	One-Dimensional Charring of LVL Under Standard Fire Exposure	72
		4.5.1 Temperature–TimeRelationship	72
		4.5.2 Charring Rate Calculation	79
	4.6	Comparison of Charring Rates with the Eurocode 5	80
	4.7	Conclusion	81
	References		82
5	**Two-Dimensional Charring Rate for Laminated Veneer Lumber from Malaysian Tropical Timber**		**83**
	5.1	Introduction	83
	5.2	Testing Material, Specimens, and Equipment	83
		5.2.1 Preparation of Specimen	84
		5.2.2 Visual Observations	85
	5.3	Determining the Residual Cross-Section: Measurement and Analysis	87
	5.4	Evaluation of Charring Rate Based on Thermocouple Readings	90
	5.5	Comparison of Charring Rates for LVL with the Eurocode 5	91
	5.6	Conclusions	92
	References		94
Index			**95**

List of Figures

Fig. 1.1	Quality of timbers; **a** large diameter logs are difficult to find, **b** the common logs size	4
Fig. 1.2	Examples of natural defects in tropical hardwood timber that can affect the strength; **a** brittle heart, **b** bark pocket, **c** hollow knot, **d** borer/grub holes	4
Fig. 1.3	Glulam as; **a** straight beam, **b** curve beam	5
Fig. 1.4	LVL board	5
Fig. 1.5	Cross-laminated timber manufacture from Malaysian hardwood timber; **a** Light Red Meranti (*Shorea spp.*), **b** White Laran (*Neolamarckia cadamba*)	5
Fig. 1.6	Application of CLT panels for load-bearing floors and walls	6
Fig. 1.7	Application of glulam in Malaysia; **a** Glulam Gallery, Johor Bahru, **b** Crop for the future research center, Semenyih, **c** Multi-purpose hall, TLDM, Lumut, **d** MITI Pavillion, Milan	8
Fig. 1.8	Schematic diagram of LVL manufacturing process	9
Fig. 1.9	LVL manufacturing process; **a** log cutting to the required length, **b** peeling process, **c** checking the quality of veneer, **d** veneer cutting, **e** veneer being dried in a conventional dryer, **f** veneer clipping, **g** veneer patching, **h** veneer arrangement by workers in resin spreading line, **i** veneer with surface gluing, **j** veneer lay-up, **k** cold pressing, **l** LVL panels after finishing	10
Fig. 1.10	LVL with layers of veneer from higher density timber	13
Fig. 1.11	Roof trusses using LVL	13
Fig. 1.12	Structural framing system using LVL	14
Fig. 2.1	Reaction to fire properties of surface products such as wall and ceiling linings	22
Fig. 2.2	Principal of grain directions (tangential, radial, and longitudinal)	23

Fig. 2.3	Diagrammatic representation of a wedge-shaped segment cut from a five-year-old hardwood tree showing the principle structural features, from Dinwoodie (2000)	24
Fig. 2.4	Part of a cellulose polymer chain, and a single glucose unit (monomer) (Lowden and Hull 2013)	25
Fig. 2.5	Fire triangle	27
Fig. 2.6	Evaporation of moisture can be seen on the surface of the wood during fire testing	28
Fig. 2.7	Moisture movement inside a burning wood sample (Bartlett et al. 2019)	29
Fig. 2.8	Combustion stages; **a** preheating the samples using a radiant heat source (heating coil), **b** the ignition of the combustible gaseous components evaporating from the sample, **c** the fire starts to spread, **d** the charred layers are observed	31
Fig. 2.9	Effect of moisture content on weight change profiles of oak during combustion at 800 °C (Orang and Tran 2015)	34
Fig. 3.1	Two main stages relevant for the fire safety in buildings in relation to building materials and structures	43
Fig. 3.2	Schematic diagram on the resistance to fire	44
Fig. 3.3	Charred samples for light red meranti after 30 min after one-dimensional fire exposure	45
Fig. 3.4	Cross-section of fire-exposed timber; **a** actual charred sample, **b** illustration of charring of wood beam, **c** charring of wood beam with temperature gradient in burning wood, illustrating the heat transfer process and internal thermal degradation	46
Fig. 3.5	Effect of temperature on the mechanical properties of softwood; **a** reduction factor for strength parallel to grain, **b** reduction factor for modulus of elasticity parallel to grain (EN1995-1-2:2004, Fig. B.4 and B.5, respectively)	47
Fig. 3.6	Reaction to fire test for glulam from Malaysian tropical timber using room corner test	48
Fig. 3.7	Graphs of reaction to fire; **a** the average heat release-rate HRRav(t) and the total heat release THR(t), **b** the FIGRA(t)-value, **c** the smoke production-rate SPRav(t) and the total smoke production TSP(t), **d** the SMOGRA(t)-value, **e** samples after reaction to fire test	51
Fig. 3.8	Changes of region of timber when subjected to fire (Cachim and Franssen 2009)	53
Fig. 3.9	Illustration on the one-dimensional fire exposure	54
Fig. 3.10	One-dimensional charring of wide cross-section (fire exposure for one side)	55
Fig. 3.11	Beam with potential for three-sided fire exposure conditions	55

List of Figures

Fig. 3.12	Effect of arris rounding on charring on the wide and narrow sides of cross-section and charring depth measurement (FireInTimber Project 2010)	56
Fig. 3.13	Charring depth for one-dimensional charring and notional charring depth (FireInTimber Project 2010)	56
Fig. 3.14	Graphic process of charring rate model of timber beam (White 2013)	57
Fig. 4.1	Charred surface after 60 min fire exposure showing the crack charcoal; **a** Dark Red Meranti, **b** Kedondong, **c** Light Red Meranti, **d** Jelutong	64
Fig. 4.2	Furnace; **a** front view, **b** fitted thermocouples to measure the temperature inside the furnace	65
Fig. 4.3	Schematic diagram of furnace cover with three slots for specimen mounting	66
Fig. 4.4	Actual furnace cover with three slots	67
Fig. 4.5	Drilling pattern and location of thermocouples	68
Fig. 4.6	Schematic diagram showing the depth positions of thermocouples	68
Fig. 4.7	Preparation of testing frame; **a** the gap filled with the fire blanket, **b** securing the specimen with G-clamp	68
Fig. 4.8	Connecting thermocouples to the electronic data logger	69
Fig. 4.9	Example of the conditions of the samples before and after fire exposure; **a** side view before fire test, **b** side view after test, **c** front view before test, **d** front view after test	71
Fig. 4.10	Comparative cross-sectional images of initial and residual specimens analyzed with CAD software Visio 2013, showing average char depths (dimensions in mm)	71
Fig. 4.11	Temperature–time relationships of Kasai	73
Fig. 4.12	Temperature–time relationships of Mengkulang	74
Fig. 4.13	Temperature–time relationships of Rubberwood	75
Fig. 4.14	Temperature–time relationships of Eucalyptus	76
Fig. 4.15	Temperature–time relationships of Kedondong	77
Fig. 4.16	Typical examples of water evaporation in samples during fire exposure; **a** Kasai, **b** Mengkulang, **c** at Thermocouples	78
Fig. 4.17	Comparison of experimental results with Eurocode 5	81
Fig. 5.1	Preparation of 100 mm width LVL; **a** gluing two pieces of 50 mm LVL, **b** the clamping process	84
Fig. 5.2	Drilling for thermocouple insertion in a two-dimensional fire test of a 100 mm wide specimen	85
Fig. 5.3	Thermocouple layout and dimensions for a 100 mm width LVL specimen; **a** layout plan, **b** Section A-A	86
Fig. 5.4	Preparation of two-dimensional fire test; **a** side view showing the 130 mm deep LVL which will be exposed to fire, **b** test frame ready for testing	86

Fig. 5.5	Example of the conditions of the samples; **a** before, **b** after fire exposure	87
Fig. 5.6	Schematic diagram of the tested samples cut into a series of blocks	89
Fig. 5.7	Residual of charred samples for Kasai LVL analyzed with Visio Professional 2013 (dimension in mm)	89
Fig. 5.8	Area for determination of char depths; **a** actual specimens, **b** method of measurement of char depth for three-sided fire exposure	89
Fig. 5.9	Determination of char depth by averaging from several points analyzed with a CAD software	90
Fig. 5.10	Temperature–time profile for Rubberwood LVL at different depths from exposed surface	91
Fig. 5.11	Comparison of experimental results with Eurocode 5	94

List of Tables

Table 1.1	Grade stress for various strength groups of structural LVL (stresses and elastic moduli expressed in N/mm^2). *Source* MS 544 Part 12	18
Table 3.1	Overview of the European reaction to fire classes for building products excluding floorings (EN 13,501–1)	49
Table 3.2	Performance criteria for reaction to fire test for building products	50
Table 3.3	Classes of reaction to fire performance for cross laminated timber products and laminated veneer lumber products for walls and ceilings	50
Table 3.4	Comparison of European and UK fire standards/classifications	52
Table 3.5	Wet and dry grade stress for various strength groups of timber (stresses and moduli expressed in N/mm^2 (*Source* Table 14, MS 544 Part 2)	59
Table 3.6	Charring rate from MS 544: Part 9	60
Table 3.7	Charring rate from BS5268: Part 4	60
Table 3.8	Design charring rates and of timber	60
Table 4.1	Timber species for LVL	64
Table 4.2	Visual observations of kedondong LVL during the fire test	70
Table 4.3	Char depth measurement of the residual cross-section	72
Table 4.4	Example calculation of charring rate for Mengkulang	79
Table 4.5	Average charring rates from experimental results (via thermocouple readings) for specimens with density greater than 480 kg/m^3 (excluding Kedondong)	80
Table 5.1	Material physical properties for two-dimensional β_n fire test for 100 mm width specimens	85
Table 5.2	Visual observations of Kasai LVL during fire test	88

Table 5.3	Thermocouples reached the 300 °C isotherm at different depths for all species in the 100 mm width LVL specimens	92
Table 5.4	Calculation of charring rate by indirect measurement for 100 mm width LVL	93
Table 5.5	Average charring rates from experimental results for LVL	93

Chapter 1
Engineered Wood Products as Construction Material

1.1 Introduction

Before the advent of rolled steel and reinforced concrete, wood served as the primary structural material in timber-rich regions worldwide (He 2023). However, recent shifts in raw material availability characterized by decreasing log diameters and a growing reliance on fast-growing plantation species have prompted a transformation in the types of timber products employed in construction. As a result, traditional solid timber products have increasingly been supplemented by Engineered Timber Products (ETPs). The term "Engineered Timber" broadly encompasses a range of products manufactured by processing raw timber logs into smaller components, which are then reassembled using adhesives. This innovative approach effectively mitigates the natural variability inherent in timber, enabling engineers to achieve enhanced strength, improved performance, and greater precision compared to conventional solid timber. Consequently, the construction industry is now capable of erecting taller, longer, and more expansive structures than ever before.

Products such as glued laminated timber (Glulam), laminated veneer lumber (LVL), and cross-laminated timber (CLT) have been utilized in various regions, including Australia, Europe, North America, and Japan, albeit in relatively small quantities for several years. However, advancements in computer technology have facilitated the production and shaping of these materials into more complex configurations, thereby enhancing manufacturing efficiency. This evolution has enabled the economically viable construction of a diverse array of structures, including residential and non-residential buildings, bridges, and industrial facilities. Today, both traditional timber products and ETPs are recognized as sustainable choices, actively promoted by governments as part of broader strategies for sustainable development and climate change mitigation (Yadav and Kumar 2022).

From a technical standpoint, modern ETPs typically offer superior and more predictable physical and mechanical properties compared to traditional timber products. These advantages include a more uniform structure, enhanced dimensional stability, and increased strength and stiffness. Initially, the development of ETPs focused on creating substitute products capable of replacing smaller dimension sawn lumber and boards as primary elements in light-frame building superstructures. However, in recent decades, there has been a notable shift toward the creation of mass timber products (MTPs). The term MTP refers to a category of EWPs characterized by larger cross-sectional dimensions, providing the construction industry with a viable alternative to structural steel and reinforced concrete (Gysling et al. 2023). This category includes thick-panel products such as CLT and structural composite lumber (SCL), as well as laminated linear elements produced through adhesive or mechanical bonding, including Glulam, nail-laminated timber (NLT), and dowel-laminated timber (DLT).

The increasing adoption of ETPs reflects a broader trend within the construction industry toward sustainable practices and innovative material solutions. As the demand for environmentally friendly building materials continues to grow, ETPs emerge as a promising option that not only meets structural requirements but also aligns with sustainability principles. Recent studies suggest that the rising demand for mass timber could have minimal impacts on forest stocks in sustainably managed forests, thereby enhancing carbon mitigation efforts through the substitution of carbon-intensive materials with mass timber (He et al. 2019). This trend underscores the potential of ETPs to contribute to a more sustainable future in construction.

Engineered wood products (EWPs), also referred to as manufactured wood or composite wood, represent a significant advancement in building materials designed to overcome the limitations of traditional solid wood. These products are created by binding or fixing layers of sawn timber, veneers, strands, or particles of wood together with adhesives, resins, or mechanical fasteners to achieve enhanced strength, durability, and versatility. The development of EWPs has emerged as a pivotal innovation in the construction industry, addressing the growing demand for sustainable and high-performance building materials. The inception of engineered timber products was driven by the need to utilize timber resources more efficiently, particularly in response to deforestation and the challenges associated with sourcing large, defect-free solid wood sections. By repurposing smaller, lower-grade, or less-desirable timber components, engineered timber products maximize the use of raw materials and contribute to sustainable construction practices.

The manufacturing process of EWPs enhances the mechanical properties of wood, resulting in materials that exhibit superior strength-to-weight ratios compared to traditional solid wood (He 2023; Yadav and Kumar 2022). The versatility of engineered wood allows for its application in diverse structural components, ranging from beams and columns to flooring systems, thereby expanding its utility in modern architecture (Gysling et al. 2023). The rise of engineered wood products is closely linked to environmental considerations. As the construction industry seeks to reduce its carbon footprint, EWPs present a viable alternative to conventional materials like concrete and steel, which are associated with higher greenhouse gas emissions during production (Yadav and Kumar 2022; He et al. 2019). The use of wood, a renewable resource, not only sequesters carbon dioxide but also promotes sustainable forestry practices (Yadav and Kumar 2022; Li et al. 2022). Furthermore, advancements in wood modification technologies have enhanced the durability and performance of engineered wood, making it suitable for a wider range of applications, including those in challenging environments (Sandberg et al. 2017).

Despite their advantages, the integration of engineered wood products into construction practices is not without challenges. The absence of standardized inspection methodologies for these products can lead to inconsistencies in quality and performance (Loferski et al. 2013). Additionally, the proprietary nature of many engineered wood products complicates the establishment of uniform installation standards across different manufacturers (Loferski et al. 2013). Nevertheless, ongoing research and development efforts aim to address these issues, ensuring that engineered wood products can meet the rigorous demands of contemporary construction while adhering to sustainability principles (Song et al. 2018).

1.2 Benefit

Timber is currently experiencing a renaissance in the building industry, a trend that warrants a thorough exploration of the underlying reasons for its resurgence and its potential applicability in other fields of engineering. This renewed interest in timber can be attributed to several interrelated factors, including the depletion of traditional timber resources, advancements in engineered wood technologies, and the growing emphasis on sustainable construction practices.

One of the primary drivers behind the increased utilization of timber in construction is the diminishing supply of wider-width and longer-length structural-grade timbers. As natural forests are subjected to overharvesting and environmental degradation, the availability of high-quality timber that meets the stringent requirements of modern construction has become increasingly limited (Fig. 1.1). This scarcity has prompted the construction industry to seek alternative solutions that can deliver the necessary structural performance while adhering to sustainability principles. In this context, engineered wood products (EWPs) such as glued laminated timber (Glulam), laminated veneer lumber (LVL), and cross-laminated timber (CLT) have emerged as viable substitutes.

Timber in its natural form, e.g., lumber or log, may not be the most efficient product for a particular load-carrying purpose. The presence of defects such as knots, holes etc. (Fig. 1.2) in a piece of lumber will severely limit its load-carrying capacity. By sawing the piece into thinner layers or laminae, or peeling it into thin sheet of veneers and gluing it with adhesive, a stronger but not necessarily stiffer member is produced. The strength of the member can be increased by either removing the knots or by dispersing the knot within the layers particularly placing it at the neutral plane.

This positioning of various grades of lamination coupled with the random dispersal of knots and other strength reducing characteristics has resulted in improved strength of laminated system over its individual laminates. The end products are more homogeneous and hence have more reliable properties. Thus, the safety factor is lower and

Fig. 1.1 Quality of timbers; **a** large diameter logs are difficult to find, **b** the common logs size

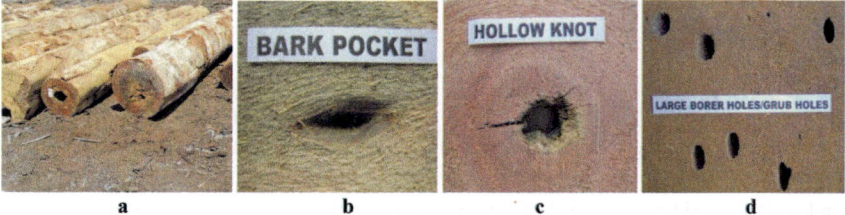

Fig. 1.2 Examples of natural defects in tropical hardwood timber that can affect the strength; **a** brittle heart, **b** bark pocket, **c** hollow knot, **d** borer/grub holes

smaller seized LVL, glulam, or CLT material can obtain compared to solid wood of the same strength.

Engineered wood products are manufactured by combining smaller pieces of timber or wood fibers using advanced adhesive technologies and manufacturing processes. This approach not only maximizes the use of available timber resources but also enhances the mechanical properties of the resulting products.

For instance, Glulam is created by bonding together layers of dimensional lumber, allowing for the production of large, strong beams that can span greater distances than traditional solid wood (Fig. 1.3). Similarly, LVL (Fig. 1.4) is produced by gluing together thin layers of wood veneers, resulting in a material that exhibits superior strength and stability, making it ideal for use in beams, headers, and other structural applications. CLT (Fig. 1.5), on the other hand, consists of multiple layers of lumber oriented at right angles to one another, providing exceptional strength and rigidity, which is particularly advantageous in multi-story construction.

The advancements in manufacturing and adhesive technologies have played a crucial role in the evolution of these engineered products. Modern adhesives are designed to provide superior bonding strength and durability, enabling engineered wood products to meet or exceed the performance characteristics of traditional solid timber. Furthermore, innovations in manufacturing processes have allowed for greater precision in the production of EWPs, resulting in materials that are not only

1.2 Benefit

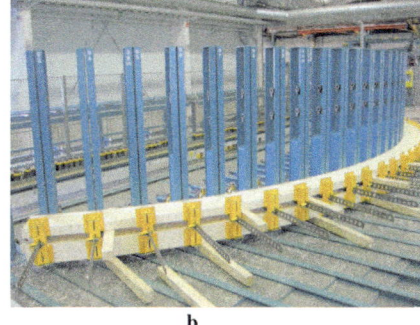

Fig. 1.3 Glulam as; **a** straight beam, **b** curve beam

Fig. 1.4 LVL board

Fig. 1.5 Cross-laminated timber manufacture from Malaysian hardwood timber; **a** Light Red Meranti (*Shorea spp.*), **b** White Laran (*Neolamarckia cadamba*)

stronger but also more uniform in quality. This consistency is essential for meeting the high standards set by the construction industry, where structural integrity and safety are paramount.

In addition to addressing the challenges posed by the depletion of traditional timber resources, the growing emphasis on sustainable construction practices has further fuelled the renaissance of timber in the building industry. As concerns about climate change and environmental degradation intensify, there is a pressing need for construction materials that minimize carbon footprints and promote resource efficiency. Timber, being a renewable resource, offers significant environmental advantages over conventional materials such as concrete and steel, which are associated with high greenhouse gas emissions during production. The use of engineered wood products not only sequesters carbon but also supports sustainable forestry practices, making it an attractive option for environmentally conscious builders and architects.

Moreover, the versatility of engineered wood products allows for their application in a wide range of engineering fields beyond traditional construction. Cross-laminated timber (CLT) may use for prefabricated wall and floor panels (Fig. 1.6). The cross-laminating manufacturing process provides improved dimensional stability to the product which allows for prefabrication of long, wide floor slabs, long single-story walls, and tall plate heights.

In Malaysia, the application of EWPs is currently limited primarily to glued laminated timber (glulam). Recognizing Malaysia's position as a timber-producing country, the Malaysian Timber Industry Board (MTIB) has taken significant steps to promote and utilize glulam within the construction industry. One notable initiative is the construction of the first iconic commercial building using glulam, the Malaysia Timber Exhibition Centre, Glulam Gallery in Mukim Tebrau, Johor Bahru. Completed in December 2011, this landmark building serves as a center to showcase and promote locally produced glulam for use in construction. Through this project, MTIB has established itself as a key advocate for the adoption of innovative engineered timber products in the domestic market. It also aims to encourage the timber-based industry to manufacture glulam for both structural and aesthetic applications, aligning with the goals outlined in the National Timber Industry Policy (NATIP). Figure 1.7a illustrates the Malaysia Timber Exhibition Centre located in Mukim

Fig. 1.6 Application of CLT panels for load-bearing floors and walls

1.3 Laminated Veneer Lumber (LVL)

Tebrau, Johor. The Glulam Gallery comprises three exhibition halls constructed using Resak and Keruing timber. The structure features 39 glulam portals, each measuring 700 mm in width and 150 mm in thickness. These portals are connected using glued-in steel rods, bolted and welded to steel brackets and plates for stability and durability. The inner side of the roof is composed of solid Keranji and Balau timber pre-coated boards, while the roof is covered with approximately 350,000 aesthetically appealing Belian shingles.

Other examples of glulam structures in Malaysia include the glulam portal frame at Semenyih, as shown in Fig. 1.7b, which was constructed using Malagangai glulam. Another significant project is the TLDM multi-purpose hall (Fig. 1.7c), which demonstrates the advantages of using glulam in buildings located near coastal areas, given its resistance to challenging environmental conditions. Additionally, Malaysia's glulam expertise has been showcased internationally, with a notable project in Italy depicted in Fig. 1.7d.

These projects highlight Malaysia's growing interest and capability in utilizing glulam as a sustainable and innovative building material, contributing to the development of the engineered timber industry both locally and globally.

The renaissance of timber in the building industry is driven by a combination of factors, including the depletion of traditional timber resources, advancements in engineered wood technologies, and the growing emphasis on sustainability. As the construction industry continues to embrace engineered wood products such as Glulam, LVL, and CLT, it is essential to recognize their potential applications in other fields of engineering. The ongoing development of these materials, coupled with a commitment to sustainable practices, positions timber as a key player in future of engineering across various disciplines.

1.3 Laminated Veneer Lumber (LVL)

Laminated veneer lumber (LVL) is a high-strength engineered wood product that has gained significant traction in the construction industry due to its superior mechanical properties and versatility. This composite material is made by bonding together thin layers of wood veneers with adhesives, resulting in a product that exhibits enhanced strength, stability, and durability compared to traditional solid wood. As the demand for sustainable and efficient building materials continues to rise, LVL stands out as a viable alternative for various structural applications especially for roof trusses.

The manufacturing process of LVL involves several key steps, which are crucial for ensuring the quality and performance of the final product. The process begins with the selection of high-quality logs, which are typically rotary peeled to produce thin veneers, usually ranging from 3 to 4 mm in thickness (Chybiński and Polus 2022; Purba et al. 2019). The veneers are then dried to a specific moisture content, which is essential for achieving optimal adhesion during the bonding process. Figure 1.8 shows the schematic diagram on the LVL manufacturing process, and Fig. 1.9 shows the actual manufacturing process for LVL.

Fig. 1.7 Application of glulam in Malaysia; **a** Glulam Gallery, Johor Bahru, **b** Crop for the future research center, Semenyih, **c** Multi-purpose hall, TLDM, Lumut, **d** MITI Pavillion, Milan

1.3 Laminated Veneer Lumber (LVL)

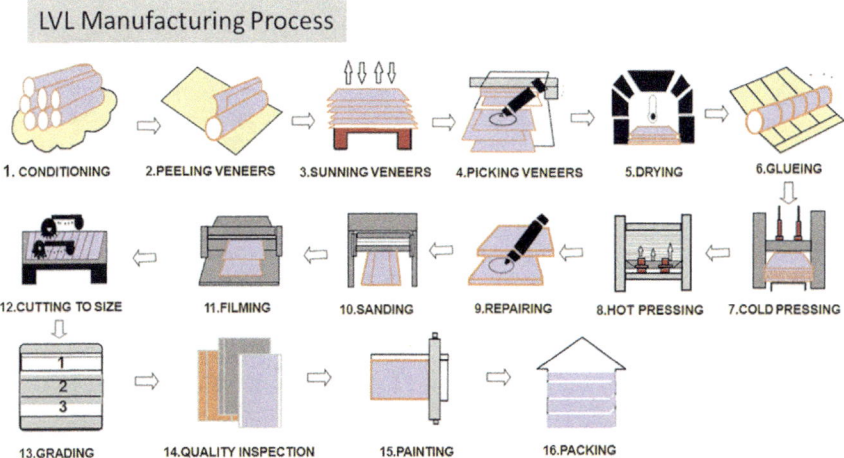

Fig. 1.8 Schematic diagram of LVL manufacturing process

Once the veneers are prepared, they are arranged in a parallel orientation, with the grain of each layer aligned in the same direction. This alignment is critical, as it allows for the efficient distribution of loads across the laminated structure, enhancing its overall strength and stiffness (Vladimirova and Gong 2022). After arranging the veneers, a suitable adhesive is applied, and the layers are pressed together under controlled conditions of temperature and pressure. Melo and Del Menezzi (2015) reported that the adhesive type and bonding quality are critical to LVL's performance, with phenol–formaldehyde resin provide greater resistance in joint analysis. This step is vital for ensuring a strong bond between the veneers, which contributes to the mechanical properties of the LVL.

The pressing process typically involves the use of hydraulic or mechanical presses, which apply uniform pressure to the assembled veneers. The duration and conditions of the pressing process can significantly influence the final properties of the LVL, including its strength and dimensional stability (Onat 2023). After pressing, the laminated panels are cured, allowing the adhesive to fully set and bond the layers together.

Once cured, the LVL panels are cut to the desired dimensions and subjected to quality control tests to ensure they meet the required standards for structural applications. These tests often include assessments of mechanical properties such as bending strength, modulus of elasticity, and shear strength (Sharapov et al. 2021; Basuki et al. 2018). The final product is then ready for use in various construction applications, including beams, columns, and trusses.

Fig. 1.9 LVL manufacturing process; **a** log cutting to the required length, **b** peeling process, **c** checking the quality of veneer, **d** veneer cutting, **e** veneer being dried in a conventional dryer, **f** veneer clipping, **g** veneer patching, **h** veneer arrangement by workers in resin spreading line, **i** veneer with surface gluing, **j** veneer lay-up, **k** cold pressing, **l** LVL panels after finishing

Fig. 1.9 (continued)

1.4 Advantages of Laminated Veneer Lumber

LVL offers several advantages over traditional solid wood and other engineered wood products. One of the primary benefits is its superior mechanical properties. Studies indicate that LVL exhibits superior performance characteristics, including resistance to splitting and enhanced load-bearing capacity, compared to traditional timber materials (Perçin and Ülker 2023; Li and Hu 2010). This makes LVL an excellent alternative for structural applications where reliability and performance are critical.

Due to the manufacturing process, LVL exhibits a higher strength-to-weight ratio, making it an ideal choice for structural applications where load-bearing capacity is critical (Purba et al. 2019; Perçin and Uzun 2023). Erdil et al. (2009) found that the strength and stiffness of LVL are significantly enhanced compared to traditional lumber due to the controlled manufacturing process, which minimizes the impact of natural wood defects. Nehdi et al. (2021) showed that LVL exhibits excellent flexural strength and stiffness, making it a viable alternative to steel and concrete in some structural elements. Bengtsson and Kliger (2003) compared LVL with other engineered wood products such as Glulam and found that LVL performed well under both short- and long-term loading, retaining its structural integrity in demanding conditions.

Another significant advantage of LVL is its dimensional stability. The manufacturing process minimizes the effects of natural wood defects, such as knots and warping, which can compromise the integrity of solid wood products. As a result, LVL maintains its shape and performance characteristics over time, even in varying environmental conditions (Alamsyah et al. 2023). This stability is particularly beneficial in applications where precise dimensions are required, such as in prefabricated construction systems. For instance, in a 15-layer LVL, the effect of a defect in one layer is only 1/15 of what it would be in a piece of solid wood of the respective dimension (Moody et al. 1999) as the impact of a defect in one layer is significantly mitigated compared to a solid wood piece of the same dimensions. This phenomenon is primarily attributed to the structural design of LVL, which allows for the distribution of loads across multiple layers. The laminated structure of LVL effectively reduces the influence of defects, as the mechanical properties are averaged out over the layers. Consequently, the effect of a defect in one layer is considerably less pronounced than in a solid wood piece, where defects can lead to a more significant reduction in strength and stability. Additionally, the uniformity of LVL allows for more predictable performance, reducing the risk of failure in structural applications.

Research has shown that the mechanical properties of LVL are less sensitive to defects than those of solid wood. For instance, Cui et al., (2024) highlights that the laminated structure of LVL helps in mitigating the effects of natural defects, thus providing enhanced mechanical performance and flexibility in size. Similarly, Atreya (2023) discusses how the layered nature of engineered wood, including LVL, allows for a more uniform distribution of stress, which diminishes the impact of localized defects. Furthermore, Soukphaxay et al. (2021) note that various factors, including the presence of defects, play a crucial role in determining the mechanical properties of LVL, reinforcing the idea that defects in a multi-layered system have a reduced overall impact compared to solid wood. Moreover, the findings of Bal and Bektaş (2012) support this assertion by demonstrating that the impact bending strength of LVL is influenced by defects, but the overall effect is less severe than in solid wood due to the composite nature of LVL (Bal and Bektaş 2012). LVL can also be manufactured using veneers from different higher density and placed it at the top and bottom layers to enhance the mechanical properties as shown in Fig. 1.10.

Laminated veneer lumber is widely used in various construction applications due to its strength, stability, and versatility which allows it to be produced in a wide range of dimensions, making it suitable for applications where traditional sawn timber may not suffice. The ability to customize the size and shape of LVL components is particularly advantageous in modern construction, where design flexibility is often required. The ability allows for the creation of components that are ideally suited for specific applications, such as large beams, column members, and flanges for proprietary prefabricated I-joists (Kurt and Çil 2012). Additionally, LVL is employed in various structural applications, including utility structures, roadway signposts, and truck decking with hardwood face veneers, showcasing its versatility and strength (Girardon et al. 2016). LVL has received type approval in several countries across Europe and North America, affirming its reliability for use in load-bearing constructions. LVL is becoming increasingly popular in domestic roof trusses (Fig. 1.11)

1.4 Advantages of Laminated Veneer Lumber

Fig. 1.10 LVL with layers of veneer from higher density timber

because of its excellent tension and compression properties parallel to the grain. LVL is also used primarily as structural framing for residential and commercial construction and is well suited to applications where open web steel joists and light steel beams might be considered (Fig. 1.12).

LVL is an environmentally friendly option, as it can be produced from fast-growing plantation species and lower-grade wood materials that may not be suitable for solid

Fig. 1.11 Roof trusses using LVL

Fig. 1.12 Structural framing system using LVL

wood products. This approach maximizes the utilization of available timber resources and supports sustainable forestry practices (Yadav and Kumar 2022). As a renewable resource, LVL contributes to reducing the carbon footprint of construction projects, aligning with the growing emphasis on sustainability in the building industry.

In addition to structural performance, the aesthetic qualities of LVL are also considered in its design. LVL can be finished to achieve an architectural-grade appearance, which is important for applications where visual appeal is a priority. Some manufacturers offer finished or architectural-grade appearances at an additional cost, which can be particularly beneficial in applications where visual appeal is paramount. When LVL is utilized in such contexts, traditional wood finishing techniques can be employed to enhance the grain and protect the wood surface, resulting in a finished appearance that closely resembles plywood or solid lumber on the beam face (Wang et al. 2015). This adaptability makes LVL suitable for a variety of architectural applications, where both functionality and aesthetics are essential. This characteristic enhances its suitability for both structural and decorative applications in Malaysian architecture (Muthuveeran et al. 2020).

In terms of construction practicality, LVL can be easily cut to length on-site, which adds to its convenience in building projects. The fastening and connection details for LVL are comparable to those of solid sawn lumber, facilitating its integration into existing construction practices. However, it is crucial that any cutting, notching, or drilling is performed in accordance with the manufacturer's specifications to maintain the structural integrity of the material (Romero et al. 2023). This adherence to guidelines ensures that the mechanical properties of LVL are preserved, allowing it to perform effectively in load-bearing applications.

Furthermore, the design of LVL must also take into account local environmental conditions, such as humidity and temperature variations, which can affect its performance. The Malaysian climate necessitates careful consideration of moisture content and treatment methods to prevent issues such as warping or decay (Taufik and Sojak 2019). Adhering to the Malaysian Standards ensures that LVL products are adequately designed to withstand these environmental challenges, thereby enhancing their longevity and reliability in construction.

Laminated veneer lumber represents a significant advancement in engineered wood products, combining the natural benefits of wood with modern manufacturing techniques. Its superior mechanical properties, dimensional stability, and environmental advantages make it a preferred choice for a wide range of construction applications. As the construction industry continues to prioritize sustainability and resource efficiency, LVL is poised to play a crucial role in future of building materials.

1.5 Laminated Veneer Lumber Design Based on Malaysian Standard

The design of laminated veneer lumber (LVL) in Malaysia is guided by specific standards that ensure its structural integrity and suitability for various applications. The Malaysian Standards (MS) provide a framework for the design and use of LVL, ensuring that it meets the necessary safety and performance criteria. For instance, the Malaysian Standard MS 544 Part 12:2001 outlines the requirements for LVL, emphasizing the importance of quality control in manufacturing processes to achieve the desired mechanical properties.

In the context of structural applications, LVL is often utilized in load-bearing constructions, where its strength and durability are critical. The design considerations for LVL include factors such as load capacity, span length, and environmental conditions. The Malaysian Standard MS 1184:2014, which pertains to the design of timber structures, provides guidelines on the allowable stress and load combinations that must be considered when designing with LVL. This standard is essential for ensuring that LVL components can safely support the anticipated loads in various construction scenarios.

1.5.1 Design Method

Structural design encompasses the selection of materials, member types, sizes, and configurations to safely carry loads while ensuring suitability for intended use. Additionally, it aims to minimize construction and maintenance costs. In the design process, it is crucial to ensure that stresses induced by loads at all critical points within a structure do not exceed the material strengths at those points. This is essential for maintaining the safety and integrity of the structure. To achieve this, structural design adheres to three primary philosophies: permissible stress design, load factor design, and limit state design.

In Malaysia, the current approach to timber design predominantly follows the permissible stress design philosophy. This method, also known as allowable stress design or elastic design, restricts the stresses that develop in a structure due to service or working loads to remain within the elastic limit of the materials used. The elastic limit is determined by applying factors of safety, which ensure that the stresses do not exceed predetermined thresholds (Abubakar et al. 2023). The permissible stress design focuses on limiting specific stresses in structural components to a fraction of the material's specified strength, with the magnitude of the safety factor varying based on the required level of safety. This method assumes that materials behave elastically under service conditions, which is critical for ensuring structural performance.

It is important to note that in the permissible stress design philosophy, partial safety factors (modification factors) are applied solely to material properties when calculating permissible stresses, rather than to the loads themselves. This approach does not provide a safety margin based on collapse loads, which can be a limitation in certain design scenarios.

The Malaysian Standard for timber design is encapsulated in MS 544: Code of Practice for Structural Use of Timber, which is tailored to different types of timber and timber composites. This standard is divided into several parts, each addressing specific aspects of timber design.

Part 1: General.
Part 2: Permissible stress design of solid timber.
Part 3: Permissible stress design of glued laminated timber.
Part 4: Timber panel products:

   Section 1: Structural plywood.
   Section 2: Marine plywood.
   Section 3: Cement bonded particle board.
   Section 4: Oriented strand board.

Part 5: Timber joints.
Part 6: Workmanship, inspection and maintenance.
Part 7: Testing.
Part 8: Design, fabrication and installation of prefabricated timber for roof trusses.
Part 9: Fire resistance of timber structures.

   Section 1: Method of calculating fire resistance of timber members.

Part 10: Preservative treatment of structural timbers.
Part 11: Recommendation for the calculation basis for span tables.

   Section 1: Domestic floor joists
   Section 2: Ceiling joists
   Section 3: Ceiling binders.
   Section 4: Domestic rafters.

Part 12: Laminated veneer lumber for structural application.

1.5.2 Design Considerations

The LVL manufactured in Malaysia must conform to the Malaysian Standard MS 2209: Structural Laminated Veneer Lumber: Performance Requirements and Minimum Manufacturing Requirements, to have structural properties appropriate for its intended application, determined by testing and evaluation methods specified in by MS 544 Part 12: Structural Laminated Veneer Lumber.

A major consideration for design of any timber structure is the determination of design values for the allowable stresses or permissible stresses and the modulus of elasticity to use for any given piece of timber. There are many factors which influence timbers strength, and hence, they should be considered in the analysis-design process of all structural timber members, assemblies, and frameworks.

The international standardization for stress grades have not been established for LVL. Rather, standard procedures are available for developing the stress grades according to established international standard such as ASTM D 5456 and AS/NZS 4357. Commonly, each manufacturer follows these procedures and submits supporting data to the appropriate regulatory authority to established stress grades and design properties for the product. Thus, design information for LVL varies among manufacturers and is given in their products' literature.

In Malaysia, the strength of structural LVL is described in terms of grade stress or basic working stress. LVL having similar grade stresses and stiffness properties has been grouped together into strength groupings (SG) for simplicity in design procedure as established in the MS 544 Part 12: Structural Laminated Veneer Lumber and shown here in Table 1.1. This is mainly due to the fact that LVL is similar to sawn timber, regardless of species and size, is still variable in strength and stiffness. Seven such strength groups are formed based on the basic working stresses, namely SG1, SG2, SG3, SG4, SG5, SG6, and SG7, in the order of decreasing stiffness and strength.

1.6 Conclusions

In summary, engineered wood products represent a significant advancement in building materials, combining the natural benefits of wood with modern engineering techniques. Their ability to provide sustainable, high-performance solutions positions them as a key component in the future of the construction industry, particularly as the sector increasingly prioritizes environmental stewardship and resource efficiency.

Table 1.1 Grade stress for various strength groups of structural LVL (stresses and elastic moduli expressed in N/mm^2). *Source* MS 544 Part 12

Strength group	[a]Bending (MOR)	Tension parallel to longitudinal axis	Shear parallel to longitudinal axis	Compression		Modulus of elasticity (MOE)	
				Parallel to longitudinal axis	Perpendicular to longitudinal axis	Mean	Minimum
SG1	26.5	15.9	2.28	22.5	3.74	18,800	14,000
SG2	18.3	11.0	1.95	18.5	3.05	16,800	12,600
SG3	15.9	9.5	1.61	14.1	2.09	14,300	10,300
SG4	13.2	7.9	1.23	11.1	1.65	11,000	7600
SG5	9.5	5.7	1.07	8.5	1.14	9100	6300
SG6	8.9	5.3	0.86	6.9	1.02	7300	5200
SG7	6.5	3.9	0.76	5.4	0.62	6600	3400

Note The grade stress is adopted from dry standard grade in MS 544 Part 2
[a]Strength edgewise

References

A. Abubakar, N. Haron, A. Alias, T. Law, Exploring quality dimensions from a construction perspective: a literature review. Jurnal Teknologi **85**(4), 133–141 (2023)

E.M. Alamsyah, A. Darwis, Y. Suhaya, S.S. Sutrisno Munawar, J. Malik, I. Sumardi, Modified grain orientation of laminated veneer lumber characteristics of three fast-growing tropical wood species. Bioresources **18**(3), 6132–6141 (2023)

AS/NZS 4357.0, *Structural Laminated Veneer Lumber Part 0: Specifications*. Australia/New Zealand Standards (2005)

ASTM D 5456–14, Standard *Specification for Evaluation of Structural Composite Lumber Products*, ASTM International (2014)

N. Atreya, Characterizing mechanical properties of layered engineered wood using guided waves and genetic algorithm. Sensors **23**(22), 9184 (2023)

B. Bal, İ. Bektaş, The effects of some factors on the impact bending strength of laminated veneer lumber. Bioresources **7**(4), (2012)

A. Basuki, A. Awaludin, B. Suhendro, S. Siswosukarto, Predicting bending creep of laminated veneer lumber (lvl) sengon (paraserianthes falcataria) beams from initial creep test data. Matec. Web Conferences **195**, 02028 (2018)

C. Bengtsson, R. Kliger, Bending creep of high-temperature dried spruce timber. Holzforschung **57**(1), 95–100 (2003)

M. Chybiński, Ł Polus, Experimental study of aluminium-timber composite bolted connections strengthened with toothed plates. Materials **15**(15), 5271 (2022). https://doi.org/10.3390/ma15155271

Z. Cui, Q. Chun, Z. Wang, Z. Wang, Q. Chun, J. Sun, Research on mechanical properties and fire XE "fire" resistance of flame- retardant laminated veneer lumber fabricated with fast-growing poplar. Eur. J. Wood Prod. **83**, 7 (2025)

Y.Z. Erdil, A. Kasal, J.I.Z.Z. Zhang, H. Efe, T. Dizel, Comparison of mechanical properties of solid wood and laminated veneer lumber fabricated from Turkish beech. Scotch Pine Lomb. Poplar, for. Prod. J. **59**(6), 55–60 (2009)

References

- S. Girardon, L. Denaud, G. Pot, I. Rahayu, Modelling the effects of wood cambial age on the effective modulus of elasticity of poplar laminated veneer lumber. Ann. for. Sci. **73**(3), 615–624 (2016)
- J. Gysling, C. Kahler, D. Soto, W. Mejías, P. Poblete, Characterization of glued-laminated timber supply in chile. in *Proceedings of World Conference on Timber Engineering (WCTE 2023)*, 19–22 June 2023, Oslo, Norway, 4483–4490 (2023)
- X. He, Two-dimensional acoustic emission source localization on layered engineered wood by machine learning: a case study of laminated veneer lumber plate structure. Struct. Health Monit. **23**(4), 2423–2442 (2023)
- S. He, C. Chen, T. Li, J. Song, X. Zhao, Y. Kuang, Y. Liu, Y. Pei, E. Hitz, W. Kong, W. Gan, B. Yang, R. Yang, L. Hu, An energy-efficient, wood-derived structural material enabled by pore structure engineering towards building efficiency. Small Methods, **4**(1), 1900747 (2019)
- R. Kurt, M. Çil, Effects of press pressure on glue line thickness and properties of laminated veneer lumber glued with melamine urea formaldehyde adhesive. BioResources **7**(3), 4341–4349 (2012)
- Q. Li, R. Dixon, P. Ciesielski, M. Himmel, B. Urbanowicz, D. Cao, H. Zhu, Multiscale engineering of wood for a sustainable future. ChemRxiv (2022). https://doi.org/10.26434/chemrxiv-2022-d75bh
- F. Li, Y. Hu, The dominant factor analysis of dynamic shear modulus of poplar LVL based on grey theory. Adv. Mater. Res. **113–116**, 811–814 (2010)
- J.R. Loferski, J.C. Bouldin, D.P. Hindman, Development of a methodology for the visual inspection of engineered wood products and metal hangers in residential construction. Adv. Mater. Res. **778**, 342–349 (2013)
- R.R. Melo, C.H.S. Del Menezzi, Influence of adhesive type on the properties of LVL made from Paricá (Schizolobium amazonicum Huber ex. Ducke) plantation trees. Drv. Ind. **66**(3), 205–212 (2015)
- R.C. Moody, R. Hernandez, J.Y. Liu, Glued structural members. in *Wood Handbook: Wood as an Engineering Material. U.S. Department of Agriculture, Forest Service, Forest Products Laboratory*, Madison, WI, chap. 11 (1999)
- MS 544–12:2006—Code of practice for structural use of timber—Part 12: 2006. Laminated Veneer Lumber For Structural Application. Department of Standards Malaysia
- A. Muthuveeran, O. Tahir, R. Ibrahim, M. Zairul, Risk management benefits and challenges in malaysia's landscape architecture project. Asian J. Behav. Stud. **5**(19), 25–43 (2020)
- M. Nehdi, Y. Zhang, X. Gao, L. Zhang, A. Suleiman, Experimental investigation on axial compression of resilient nail-cross-laminated timber panels. Sustainability **13**(20), 11257 (2021)
- S.M. Onat, Optimization of production parameters of densified laminated veneer lumber produced by using urea-formaldehyde resin. Drvna Industrija **74**(3), 327–335 (2023)
- O. Perçin, O. Ülker, Influence of carbon fibre layers on the strength of thermally modified laminated veneer lumber. Polímeros **33**(1), e20230007 (2023)
- O. Perçin, O. Uzun, Physical and mechanical properties of laminated wood made from heat XE "heat" -treated scotch pine reinforced with carbon fiber. BioResources **18**(3), 5146–5164 (2023)
- C. Purba, G. Pot, J. Viguier, J. Ruelle, L. Denaud, The influence of veneer thickness and knot proportion on the mechanical properties of laminated veneer lumber (LVL) made from secondary quality hardwood. Eur. J. Wood Wood Prod. **77**(3), 393–404 (2019)
- A. Romero, J. Yang, F. Hanus, H. Degée, C. Odenbreit, Push-out tests on connections for demountable and reusable steel-timber composite beam and flooring systems, in *Proceedings of World Conference on Timber Engineering (WCTE 2023)*, pp. 19–22 June 2023, Oslo, Norway, 3568–3574 (2023)
- D. Sandberg, A. Kutnar, G. Mantanis, Wood modification technologies—a review. Iforest—Biogeosciences for. **10**(6), 895–908 (2017)
- E. Sharapov, C. Brischke, S. Bicke, J. Steeg, H. Militz, Evaluation of white rot decay in phenol-formaldehyde resin treated European beech (fagus sylvatica l.) LVL by drilling resistance measurements. Eur. J. Wood Wood Prod. **80**(2), 439–449 (2021)

J. Song, C. Chen, S. Zhu, M. Zhu, J. Dai, U. Ray, Y. Li, Y. Kuang, Y. Li, N. Quispe, Y. Yao, A. Gong, U.H. Leiste, H.A. Bruck, J.Y. Zhu, A. Vellore, H. Li, M.L. Minus, Z. Jia, A. Martini, T. Li, L. Hu, Processing bulk natural wood into a high-performance structural material. Nature **554**, 224–228 (2018)

K. Soukphaxay, K. Phonetip, L. Boupha, L. Saetern, K. Khammanivong, Y. Fu, Mechanical properties assessment of laminated veneer lumber from teak plantation in laos. Walailak J. Sci. Technol. (Wjst) **18**(2), 6539 (2021)

F. Taufik, S. Sojak, Mosque architectural timeline in malaysia: from vernacular to contemporary, in *Carving The Future Built Environment: Environmental, Economic and Social Resilience*, eds. by P.A.J. Wahid, P.I.D.A. Aziz Abdul Samad, P.D.S. Sheikh Ahmad, A.P.D.P. Pujinda, 2. European Proceedings of Multidisciplinary Sciences, pp. 179–188 (2019)

E. Vladimirova, M. Gong, Veneer-based engineered wood products in construction, engineered wood products for construction. IntechOpen **28**, 2022 (2022)

J. Wang, X. Guo, W. Zhong, W. Hui-yun, P. Cao, Evaluation of mechanical properties of reinforced poplar laminated veneer lumber. BioResources **10**(4), 7455–7465 (2015)

R. Yadav, J. Kumar, Engineered wood products as a sustainable construction material: a review. Eng. Wood Prod. Constr. IntechOpen (2022)

Chapter 2
Fundamentals of Fire and Combustion

2.1 Introduction

The reaction to fire of a material refers to its behavior and contribution to the development of a fire during its initial stages. This classification encompasses several critical parameters, including ignition, heat release rate, fire spread and growth, as well as smoke production. These factors are essential in assessing how a material may influence the dynamics of a fire, particularly in its early development phase which is the stage when combustible products may contribute to the fire. The schematic diagram illustrated in Fig. 2.1 provides a visual representation of these properties and their interrelationships. It is commonly used in fire safety assessments to determine how different materials contribute to fire development. Building materials, textiles, and insulation are tested under standardized conditions to classify their reaction to fire. For example, materials with high fire resistance, such as gypsum board or treated wood, have a slower ignition time and produce lower heat release rates, reducing the risk of rapid fire spread (Östman et al. 2010).

Ignition is the first and perhaps the most crucial parameter in the reaction to fire. It denotes the point at which a material begins to combust, which can be influenced by various factors such as the material's chemical composition, moisture content, and environmental conditions. For instance, materials with lower ignition temperatures are more susceptible to catching fire, thereby contributing to the fire's initiation (Dudiak and Dzurenda 2021).

The heat release rate (HRR) is another vital parameter that quantifies the amount of heat energy released per unit of time during combustion. A higher HRR indicates that a material can contribute significantly to the intensity and severity of a fire. This parameter is critical for understanding how quickly a fire can escalate and how much energy is available to sustain the combustion process (Yang et al. 2022). Research has shown that materials with high HRR can lead to rapid fire growth, posing increased risks to life and property (Cheng et al. 2022).

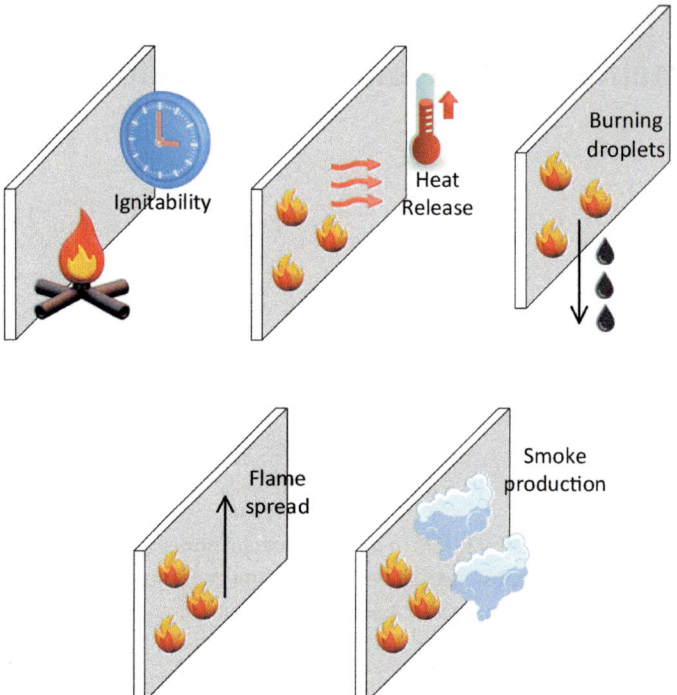

Fig. 2.1 Reaction to fire properties of surface products such as wall and ceiling linings

Fire spread and growth are closely related to both ignition and HRR. Once a fire has ignited, the rate at which it spreads can be influenced by the combustibility of surrounding materials, their arrangement, and environmental factors such as airflow. The growth of a fire is often characterized by the transition from smoldering to flaming combustion, which can significantly affect the overall fire dynamics (Qiao et al. 2024). Studies have demonstrated that the arrangement of combustible materials can either facilitate or hinder fire spread, making it a critical consideration in fire safety design (Silva et al. 2022).

Smoke production is another important aspect of fire reaction, as it can significantly impact visibility and air quality during a fire event. The composition and quantity of smoke produced are influenced by the type of material burning and the combustion conditions. For example, materials that produce dense, toxic smoke can pose additional hazards to occupants and firefighters (Orang and Tran 2015). Understanding smoke production is essential for developing effective fire suppression strategies and ensuring the safety of individuals in the vicinity of a fire (Chen et al. 2016).

Wood combustion is a chemical process in which a fuel reacts with oxygen to produce heat, light, and combustion by-products and occurs four stages, preheating, pyrolysis, gas combustion and oxidation as well as charcoal combustion and ash formation. Wood combustion is a complex chemical and physical process that involves the conversion of solid wood into heat, gases, and ash during fire exposure. This transformation occurs through multiple stages, each governed by thermal decomposition, oxidation reactions, and the influence of external conditions such as temperature, oxygen supply, and moisture content.

While reaction to fire is used to assess fire risks and material safety, combustion describes the fundamental energy-releasing process of burning. Understanding both concepts is essential for fire prevention, material selection, and optimizing combustion systems for energy efficiency.

Wood is the largest and most complex lignocellulosic material, exhibiting a wide range of physical and mechanical properties across different species. Wood is primarily composed of cellulose, hemicellulose, and lignin, which contribute to its flammability and thermal degradation properties. The thermal degradation of wood begins at approximately 200 °C, leading to pyrolysis, where wood decomposes into gases, vapors, and char (Mensah et al. 2022). Understanding these processes is crucial for predicting the fire behavior of wood.

2.2 Composition of Wood

Wood is classified as an anisotropic material due to the variation in its structure depending on the direction of the material. This means that its mechanical and physical properties differ along different orientations. The surface of a wooden beam or column can be categorized based on its orientation relative to the grain. It can either be parallel to the grain, tangential to the grain, or radial to the grain. The longitudinal axis is parallel to the grain and the radial and tangential directions are therefore perpendicular to the grain. Figure 2.2 illustrates the three primary reference planes in wood structure: the transverse surface, the radial surface, and the tangential surface.

The cross-sectional macro-structure of a tree is illustrated in Fig. 2.3, adapted from Dinwoodie (2000). Structurally, wood is composed of elongated cells that are arranged either axially (along the grain) or radially (perpendicular to the grain). These cellular arrangements contribute to the unique characteristics of wood, influencing

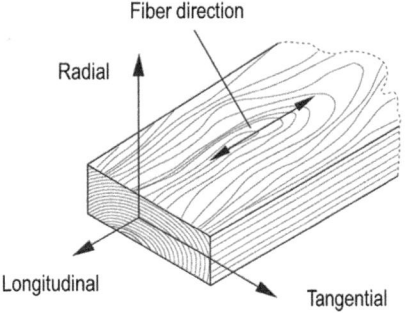

Fig. 2.2 Principal of grain directions (tangential, radial, and longitudinal)

its strength, durability, and response to external forces. In Fig. 2.3, the longitudinal axis runs parallel to the grain, while the radial and tangential directions are oriented perpendicular to the grain. A notable distinction is made between the central wood of a tree, known as heartwood, and the wood located closer to the bark, referred to as sapwood. Heartwood is typically darker in color and exhibits a higher degree of impermeability to moisture compared to the adjacent sapwood. This characteristic renders heartwood more challenging to treat with liquid-based wood preservatives; however, it is inherently more resistant to decay and fungal attacks, as noted by Kollman and Cote (1968).

The molecular structure of wood is characterized by a highly ordered crystalline cellulose, which is a large polysaccharide composed entirely of glucose units linked by β-glycosidic bonds (Purves et al. 1995). Cellulose, hemicellulose and lignin are the three polymeric materials that constitute the wood cells. Cellulose is a polymer with alternating repeat units of glucose (Fig. 2.4). The large number of hydroxyl groups on the sugar molecule, which leaves the polymer as water molecules during decomposition, result in char formation.

The wood structure is made of 70% cellulose, 12–24% lignin, and up to 1% ash forming materials. Cellulose forms the cell walls, and provides the tensile strength of the wood matrix. Hemicelluloses grow around the cellulose fibers and are a group of non-structural, low molecular weight, mostly heterogeneous polysaccharides. Lignin acts as a binding agent, uniting the cellulose and hemicelluloses, thereby enhancing the overall mechanical strength of the wood (Winandy and Rowell 1984)

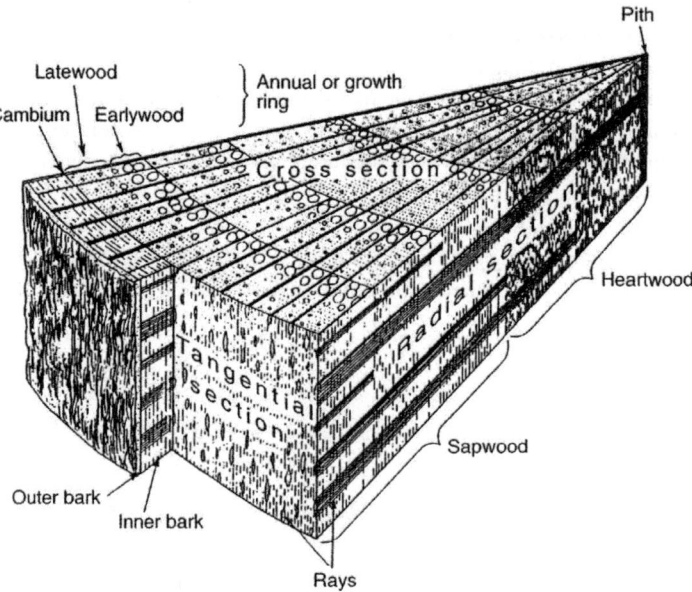

Fig. 2.3 Diagrammatic representation of a wedge-shaped segment cut from a five-year-old hardwood tree showing the principle structural features, from Dinwoodie (2000)

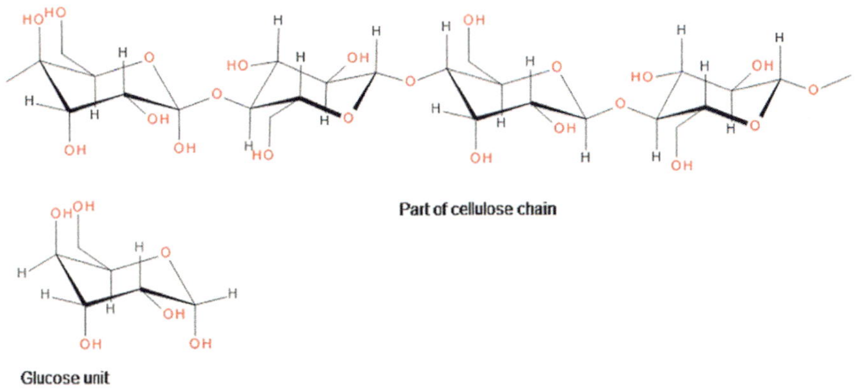

Fig. 2.4 Part of a cellulose polymer chain, and a single glucose unit (monomer) (Lowden and Hull 2013)

which allows trees to grow upright by imparting rigidity to the wood. The cellulose molecules exhibit considerable stiffness along their chain axis and are arranged in a helical orientation. Lignin plays a crucial role in stabilizing the cellulose molecules, forming microfibrils that contribute to the rigidity of the woody structure. According to Salmén (2001), the elastic modulus of cellulose is approximately 134 GPa.

Recent studies have highlighted the distinct mechanical roles of these components within wood. Cellulose, the primary and most essential constituent, is responsible for tensile strength, while lignin forms a matrix that enhances compressive resistance (Ayanleye et al. 2022). In terms of weight distribution, cellulose constitutes a major portion in both hardwoods and softwoods (Azmi 2019).

Understanding the molecular composition of wood is crucial for evaluating its reaction to fire. The crystalline structure of cellulose provides mechanical strength but also plays a significant role in how wood behaves when exposed to high temperatures.

2.3 Fire Behavior of Wood: A Molecular Perspective

The structural elements of wood namely cellulose, hemicellulose, and lignin, each play a crucial role in determining how it behaves when exposed to high temperatures. These cellulosic materials suffer from dehydration, depolymerization, and thermal decomposition in wood combustion (Zao et al. 2015, Tam et al. 2017). Understanding these components provides valuable insight into wood's thermal degradation, combustion process, and fire resistance.

(a) Cellulose

Cellulose, the primary component of wood, forms a crystalline structure that provides mechanical strength. However, this same structure also influences how wood reacts to heat. During a fire, cellulose begins to thermally degrade at relatively low temperatures, initiating a process known as pyrolysis, the chemical breakdown of wood in the absence of oxygen. This reaction releases volatile compounds that contribute to flame formation and sustained combustion.

The thermal degradation of cellulose typically occurs between 275 and 350 °C, producing unstable gases, tar-like substances (such as levoglucosan), and carbonaceous char (Lowden and Hull 2013). The rapid decomposition of cellulose plays a significant role in the ignition phase of wood burning, releasing combustible gases that feed the fire.

(b) Hemicellulose

Before cellulose begins to break down, hemicellulose, a less stable polymer, decomposes at even lower temperatures, typically between 180 and 350 °C (Östman et al. 2017). This early-stage degradation contributes to the initial release of flammable gases, making hemicellulose a key factor in determining the ease of ignition. Due to its lower thermal resistance, hemicellulose burns quickly, accelerating the fire's spread.

(c) Lignin

Unlike hemicellulose and cellulose, lignin is more thermally stable and decomposes over a wider temperature range of 250–500 °C. Lignin contributes to char formation, which plays a crucial role in the fire resistance of wood. As lignin degrades, it produces aromatic compounds and char, the latter of which acts as a protective barrier. This char layer insulates the underlying material from heat, slowing the rate of thermal degradation and reducing the overall combustibility of wood. The presence of lignin, therefore, adds a degree of fire resistance, as the formation of char can limit the spread of flames and heat penetration.

When exposed to high temperatures, wood undergoes a series of transformations that significantly alter its physical, thermal, and mechanical properties (Buchanan 2002). At temperatures exceeding 500 °C, the combined degradation of cellulose, hemicellulose, and lignin results in the production of unstable gases, tar, and carbonaceous char (Lowden and Hull 2013). These by-products play a vital role in the combustion process, influencing both flame intensity and burn duration.

The formation of char is particularly important in assessing the fire performance of wood. A well-developed char layer slows down heat transfer and limits further degradation, enhancing fire resistance. However, if the char layer is compromised through mechanical damage or exposure to prolonged high temperatures, combustion can continue, leading to further deterioration of the material.

The interplay between cellulose, hemicellulose, and lignin not only defines the structural integrity of wood but also significantly impacts its reaction to fire. While hemicellulose and cellulose contribute to the rapid ignition and burning of wood, lignin aids in char formation, influencing fire resistance. Understanding these molecular interactions provides essential knowledge for fire safety engineering, allowing for better fire-resistant treatments and improved construction practices. By manipulating these properties, we can enhance the safety and performance of wood in fire-prone environments.

Fig. 2.5 Fire triangle

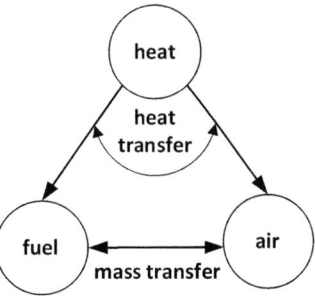

2.4 The Chemistry of Combustion

The understanding of the combustion of any material is encapsulated in the fire triangle (Fig. 2.5). This model effectively highlights the essential components that sustain a fire and how their control can lead to its extinguishment. Fire requires three elements; fuel, oxygen, and heat, to ignite and continue burning, with their proper mixing being equally important. A fire can be extinguished by removing the fuel, cutting off the oxygen supply (smothering), or lowering the temperature (cooling with water).

Combustion is a chemical reaction that typically involves a fuel, an oxidizer (usually oxygen), and heat. The basic equation for combustion can be represented as:

$$CH_4 + O_2 \rightarrow CO_2 + H_2O \tag{2.1}$$

During combustion, the fuel undergoes a series of reactions that result in the release of energy. The heat generated can initiate further combustion of nearby fuel, leading to a self-sustaining fire. Lyon et al. (2013) emphasize that the heat release rate (HRR) is a critical parameter in understanding fire behavior, as it quantifies the energy output of a fire and influences the spread and intensity of flames.

2.5 Phases of Combustion

Wood combustion occurs in four distinct stages, each characterized by different thermal and chemical processes as follows.

(a) Drying

The initial stage of combustion, known as the heating and evaporation phase, occurs when heat is applied to a piece of wood in the presence of air. This stage is essential for preparing the wood for ignition, as several key processes must take place before combustion can effectively begin. A primary function of this phase is the removal of moisture from the wood, which is crucial for efficient burning. The moisture content of wood plays a significant role in its combustion behavior, making it necessary to evaporate any retained water before ignition can proceed. This phase is particularly important because moisture acts as a heat sink; as long as water remains within the wood, the applied heat energy is primarily used to convert the water into steam rather than increasing the wood's temperature. As a result, wood with a high moisture content is more difficult to ignite, as it requires additional energy to eliminate moisture before combustion can start. Studies have shown that moisture content directly affects combustion efficiency, with higher moisture levels leading to incomplete combustion and increased emissions (Silva et al. 2022; Orang and Tran 2015).

As heat is applied, the temperature of the wood surface begins to rise, penetrating several millimeters into the material. When the surface temperature reaches approximately 100 °C, the moisture contained within the wood starts to boil and evaporate as shown in Fig. 2.6.

Wood is a hygroscopic material, meaning it can absorb and release moisture from its surroundings. While some water vapor escapes through the surface of the material during combustion, a portion migrates away from the heat source and penetrates deeper into the material, where it recondenses. This phenomenon leads to the formation of three distinct zones within the material, as illustrated in Fig. 2.7. The first zone is the dry zone, where pyrolysis occurs, characterized by the thermal decomposition of the material without the presence of moisture. The second zone is the dehydration zone, where moisture is expelled from the material as it heats up. Finally, the wet zone remains saturated with moisture, which has not yet evaporated. This stratification of zones plays a crucial role in the combustion process, influencing the efficiency and characteristics of the fire (Uysal 2005).

The dry zone is critical for pyrolysis, as it is here that the organic components of the wood begin to break down into volatile gases and char. The absence of moisture in this zone allows for the efficient release of flammable gases, which are essential

Fig. 2.6 Evaporation of moisture can be seen on the surface of the wood during fire testing

2.5 Phases of Combustion

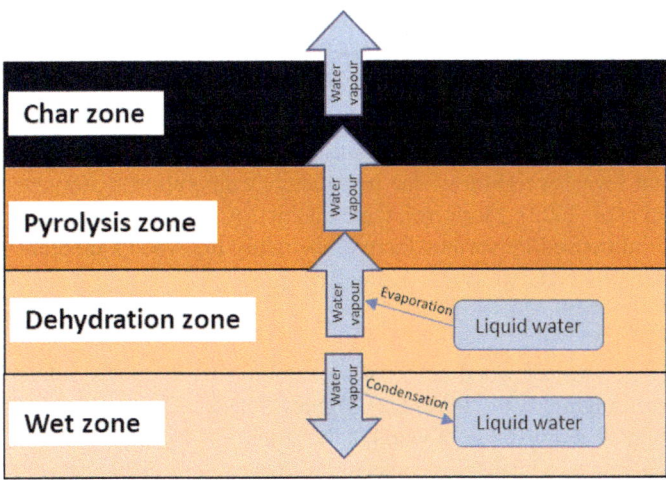

Fig. 2.7 Moisture movement inside a burning wood sample (Bartlett et al. 2019)

for sustaining combustion. Research indicates that the efficiency of pyrolysis can be significantly affected by the moisture content of the wood, as higher moisture levels can lead to lower temperatures and incomplete pyrolysis (Kasymov and Agafontsev 2017).

In the dehydration zone, the material continues to lose moisture, which is vital for preparing the wood for combustion. The rate of moisture loss can vary based on the wood species, its initial moisture content, and the intensity of the heat applied. For instance, hardwoods, which typically have lower moisture content than softwoods, may transition through this zone more quickly, allowing for a more efficient combustion process (Eriksson et al. 2014).

The wet zone, on the other hand, retains moisture that can hinder combustion. If the moisture content is too high, it can absorb heat energy that would otherwise contribute to the ignition and burning of the wood. This phenomenon can lead to incomplete combustion, resulting in the production of harmful emissions such as carbon monoxide and unburned hydrocarbons (Chandrasekaran et al. 2013). Understanding the dynamics of moisture migration and the formation of these zones is essential for optimizing combustion processes and improving the efficiency of wood as a fuel source.

Moreover, the interaction between these zones can influence the overall combustion characteristics, including heat release rates and emissions profiles. For example, studies have shown that the presence of moisture can affect the combustion temperature and the formation of particulate matter during burning (Wang et al. 2013). Therefore, effective management of moisture content and understanding the implications of these zones can lead to improved fire safety and enhanced energy efficiency in wood combustion applications.

It is important to note that no significant energy is released during this drying stage; instead, the energy is consumed in the process of evaporating the moisture. This consumption of energy is crucial as it prepares the wood for the subsequent stages of combustion, where energy release occurs. Studies have shown that the thermal degradation of wood begins with dehydration, which is evident in thermogravimetric analyses that reveal weight loss due to moisture evaporation at lower temperature ranges (Gao et al. 2003). Moreover, the dynamics of moisture removal are complex and can be influenced by various factors, including the wood's inherent properties and the environmental conditions during drying.

(b) Pyrolysis

The second stage of wood combustion is known as pyrolysis, which occurs at temperatures ranging from 200 to 600 degrees Celsius. This stage is characterized by the thermal decomposition of wood in the absence of oxygen, a critical condition that distinguishes pyrolysis from combustion (Aseeva et al. 2014). During pyrolysis, the intense heat causes the timber components to break down into volatile gases, tar, and reactive carbonaceous char. This stage not only contributes to the flames seen during combustion but also has implications for energy production and environmental sustainability through the potential generation of bio-char. This characteristic can be seen from cone calorimeter test in Fig. 2.8.

This temperature varies depending on the type of fuel and its physical state. During pyrolysis, the complex organic polymers found in wood namely cellulose, hemicellulose, and lignin, undergo significant structural changes. As the temperature rises, these components break down into a variety of products, including volatile gases (Fig. 2.8b), tar, and solid charcoal. This process, known as pyrolysis, occurs beneath the char layer (Lowden and Hall 2013). The pyrolysis zone is only a few millimeters thick and has a lower temperature compared to the char layer, which can reach temperatures of up to 800 °C (Lowden and Hall 2013).

The pyrolysis process is particularly important because it generates a range of flammable gases, including carbon monoxide (CO), hydrogen (H_2), methane (CH_4), and various hydrocarbons. These gases are crucial for sustaining combustion, as they can ignite and contribute to the overall energy output of the burning wood. Research has shown that the composition of the volatile gases produced during pyrolysis can vary significantly depending on the type of wood and the specific conditions of the pyrolysis process, such as temperature and heating rate (Dudiak and Dzurenda 2021; Gao et al. 2003). For instance, higher temperatures tend to favor the production of lighter hydrocarbons and gases, while lower temperatures may yield more tar and heavier organic compounds (Sokolovskyy et al. 2021).

Moreover, pyrolysis is responsible for the characteristic flames observed during the combustion of wood. The visible flames (Fig. 2.6c) are primarily due to the combustion of the volatile gases released during this stage. As these gases mix with oxygen in the air and ignite, they produce the bright flames that are often associated with burning wood. The color and intensity of the flames can vary based on the composition of the gases and the combustion conditions, providing insights into the

2.5 Phases of Combustion

Fig. 2.8 Combustion stages; **a** preheating the samples using a radiant heat source (heating coil), **b** the ignition of the combustible gaseous components evaporating from the sample, **c** the fire starts to spread, **d** the charred layers are observed

efficiency of the pyrolysis and subsequent combustion processes (Ivanova and Bulba 2018).

(c) Gas Combustion and Oxidation

The third stage of wood combustion, known as gas combustion and oxidation, occurs at elevated temperatures ranging from 600 to 1100° Celsius. This stage is characterized by the ignition of the volatile gases produced during the pyrolysis phase, which mix with oxygen from the surrounding environment. This stage produces the highest heat output and visible flames, while complete combustion results in the formation of carbon dioxide and water vapor. The combustion of these gases is a highly exothermic reaction, resulting in the release of significant amounts of heat and the production of visible flames. The flames observed during this stage are primarily due to the combustion of the volatile organic compounds that were released during pyrolysis, and they are indicative of the efficiency of the combustion process (Zang et al. 2022).

During gas combustion, the complete oxidation of the volatile gases results in the formation of carbon dioxide (CO_2) and water vapor (H_2O), which are the primary products of complete combustion. This process is essential for maximizing energy output and minimizing the production of harmful emissions. Studies on combustion efficiency highlight the importance of maintaining optimal oxygen levels to reduce emissions and enhance energy output (Turns 2012) and achieving complete combustion is crucial for efficient energy conversion, as it ensures that the maximum amount of energy stored in the fuel is released (Gao et al. 2003; Akagi et al. 2010).

However, the efficiency of gas combustion is highly dependent on the availability of oxygen. Insufficient oxygen can lead to incomplete combustion, which results in the formation of pollutants such as carbon monoxide (CO) and unburned hydrocarbons. Carbon monoxide is a colorless, odorless gas that poses significant health risks and contributes to air pollution. Incomplete combustion can also lead to the production of particulate matter, which can have detrimental effects on both human health and the environment (Silva et al. 2022).

To optimize combustion efficiency and minimize emissions, it is essential to maintain appropriate oxygen levels during this stage. Various combustion technologies, such as staged combustion and the use of secondary air, have been developed to enhance oxygen availability and promote complete combustion (Orang and Tran 2015; Chen et al. 2016). Additionally, advancements in combustion modeling and monitoring techniques have improved our understanding of the combustion process, allowing for better control of combustion conditions and emissions (Wang and Tian 2019).

(d) Charcoal Combustion

The fourth stage of wood combustion is characterized by the combustion of charcoal, which occurs at temperatures ranging from 800 to 1300° Celsius (1472 to 2372 degrees Fahrenheit). After the volatile gases produced during the pyrolysis stage have been burned off, a solid residue known as charcoal remains. Charcoal is primarily composed of carbon, along with smaller amounts of other elements such as hydrogen and oxygen. Due to its high carbon content, charcoal burns slowly and provides a steady heat output, making it a preferred fuel source for various applications, including cooking and heating (Austine et al. 2022; Singh et al. 2022). The burning of charcoal primarily results in the formation of carbon dioxide (CO_2) when oxygen supply is sufficient, along with small amounts of carbon monoxide (CO) in cases of incomplete combustion (Antal and Grønli 2003).

The combustion of charcoal is notable for producing minimal smoke compared to the burning of raw wood. This is largely due to the reduced presence of volatile organic compounds in charcoal, which have already been released during the earlier pyrolysis stage. As a result, the combustion of charcoal is often associated with cleaner burning and lower emissions of particulate matter and other pollutants (Ivanova and Bulba 2018). Research into modern pyrolysis and carbonization techniques continues to enhance the quality and efficiency of charcoal production, ensuring sustainable use of this valuable fuel source (Bridgwater 2012).

The efficiency of charcoal combustion can be influenced by several factors, including temperature, airflow, and the physical characteristics of the charcoal itself, such as its density and surface area. Optimizing these parameters can lead

2.6 Factors Affecting Wood Combustion Efficiency

In general, the thermal deterioration of wood and wood products is being researched for two reasons: wood may be utilized as a source of energy and chemicals, and it can also contribute to fire propagation in a building (Lee et al. 2011).

The combustion of wood is influenced by several factors that determine the efficiency, heat output, and emissions of the burning process. These factors include the moisture content of the wood, its chemical composition, the availability of oxygen, temperature conditions, and the size and arrangement of the wood pieces.

One of the most significant factors affecting wood combustion is moisture content. Studies have shown that the presence of moisture can lead to incomplete combustion, resulting in the production of harmful emissions such as carbon monoxide (CO) and unburned hydrocarbons (Galgano et al. 2013). Moreover, the volatile pyrolysis products generated during the combustion process are heavily influenced by the moisture content, as they represent a significant portion of the energy available from the wood (Galgano et al. 2013). Freshly cut or "green" wood contains a high percentage of water, often exceeding 50% by weight. High moisture content requires additional energy to evaporate the water before combustion can proceed efficiently, leading to lower heat output and increased smoke production. In contrast, well-seasoned wood, with a moisture content below 20%, burns more efficiently and produces fewer pollutants (Simoneit et al. 2002). Research indicates that the moisture content in alder wood can decrease by as much as 9.1% due to the evaporation of water into a saturated steam environment during thermal treatment, highlighting the efficiency of heat transfer in this process (Dudiak and Dzurenda 2021).

Another critical factor is the chemical composition of the wood, which varies depending on the species. Hardwoods, such as oak and maple, have a higher density, tightly packed fiber structure and energy content per unit volume, resulting in a longer, more sustained burn. Softwoods, such as pine and fir, contain higher levels of volatile compounds and resins, leading to a faster and sometimes more intense combustion process. The variation in combustion behavior between hardwoods and softwoods becomes more pronounced in samples with higher moisture content, as illustrated in Fig. 2.9. When examining the weight profile of different wood species with a 40% moisture content, the reduction in the devolatilization rate is more significant in low-density softwood. This phenomenon can be attributed to the dense structure of hardwoods, which restricts the movement of water vapor. As a result, heat takes longer to penetrate the material, delaying the drying and combustion process (Orang and Tran 2015).

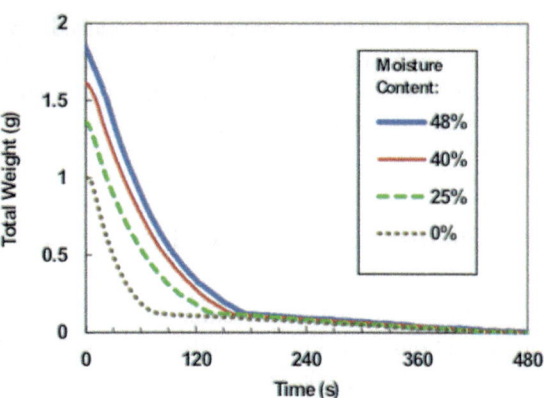

Fig. 2.9 Effect of moisture content on weight change profiles of oak during combustion at 800 °C (Orang and Tran 2015)

Additionally, the presence of lignin, cellulose, and hemicellulose affects the burning characteristics, with lignin contributing to higher energy output due to its complex molecular structure (White 1987). The presence of incombustible minerals, such as calcium carbonate, can render wood less flammable and alter its combustion behavior (Merk et al. 2016).

Combustion conditions, including temperature and airflow, are vital for achieving complete combustion. Higher combustion temperatures generally promote more complete combustion, leading to increased energy output and reduced emissions. Conversely, insufficient oxygen supply can result in incomplete combustion, producing pollutants such as such as carbon monoxide (CO), particulate matter, and unburned hydrocarbons (Johansson et al. 2004). Research indicates that the combustion dynamics, including the transition between flaming and smoldering phases, significantly influence the emissions profile and the efficiency of the combustion process (Haslett et al. 2018; Heringa et al. 2012).

Temperature also plays a crucial role in wood combustion. The process occurs in different stages, starting with pyrolysis, where heat causes the release of volatile gases. These gases ignite when temperatures exceed approximately 300 °C (572°F), producing flames and releasing the majority of the wood's energy. In the later stages, charcoal combustion occurs at even higher temperatures, sustaining heat output until only ash remains. Maintaining high combustion temperatures ensures that volatile compounds are burned completely, minimizing emissions and increasing thermal efficiency (McDonald et al. 2000).

2.7 Summary

The study of fire and combustion is a multi-disciplinary field that integrates principles from chemistry, physics, and engineering. As fire hazards continue to evolve with the introduction of new materials and technologies, ongoing research and innovation will be vital in enhancing fire safety measures and improving our understanding of combustion dynamics. By leveraging this knowledge, we can better protect lives, property, and the environment from the devastating effects of fire.

References

S.K. Akagi, R.J. Yokelson, C. Wiedinmyer, M.J. Alvarado, J.S. Reid, T. Karl, J.D. Crounse, P.O. Wennberg, Emission factors for open and domestic biomass burning XE "burning" for use in atmospheric models. Atmos. Chem. Phys. Discuss. **10**, 27523–27602 (2010)

M.J. Antal, M. Grønli, The art, science, and technology of charcoal production. Ind. Eng. Chem. Res. **42**(8), 1619–1640 (2003)

R. Aseeva, B. Serkov, A. Sivenkov, *Fire Behavior and Fire XE "fire" Protection in Timber Buildings* (Springer, Dordrecht, IX, 2014), p.290

O. Austine, A.O. Otieno, P.G. Home, J.M. Raude, S.I. Murunga, A. Gachanja, Heating and emission characteristics from combustion of charcoal and co-combustion of charcoal with faecal char-sawdust char briquettes in a ceramic cook stove. Heliyon **8**(8), e10272 (2022)

S. Ayanleye, K. Udele, V. Nasir, X. Zhang, H. Militz, Durability and protection of mass timber XE "mass timber" structures: a review. J. Build. Eng. **46**, 103731 (2022)

A.I. Bartlett, R.M. Hadden, L.A. Bisby, A review of factors affecting the burning behaviour of wood for application to tall timber construction. In Fire Technol. **55**(1). Springer New York LLC (2019)

A.H. Buchanan, *Structural Design for Fire XE "fire" Safety* (Wiley, Chichester, UK, 2002)

S.R. Chandrasekaran, P.K. Hopke, A. Hurlbut, M. Newtown, Characterization of emissions from grass pellet combustion. Energy Fuels **27**(9), 5298–5306 (2013)

C. Chen, R. Alén, J. Lehto, H. Pakkanen, Combustion properties of spruce black liquor droplets: sulfur-free pulping and influence of hot-water pretreatment. Nord. Pulp Pap. Res. J. **31**(4), 531–539 (2016)

X. Cheng, D. Lu, K. Yue, W. Lu, Z. Zhang, Fire resistance improvement of fast-growing poplar wood based on combined modification using resin impregnation and compression. Polymers (Basel) **14**(17), 3574 (2022)

J.M. Dinwoodie, *Timber Its Nature and Behaviour*, 2nd edn. (E & FN Spon, London, 2000)

M. Dudiak, L. Dzurenda, Changes in the physical and chemical properties of alder wood in the process of thermal treatment with saturated water steam. Coatings **11**(8), 898 (2021)

A.C. Eriksson, E.Z. Nordin, R. Nyström, E. Pettersson, E. Swietlicki, C. Bergvall, R. Westerholm, C. Boman, J.H. Pagels, Particulate PAH emissions from residential biomass combustion: time-resolved analysis with aerosol mass spectrometry. Environ. Sci. Technol. **48**(12), 7143–7150 (2014)

A. Azmi, *Development of Characteristic Value for Compressive Strength Properties of Selected Malaysian Tropical Hardwood Timber*. Ph.D. Thesis (2019)

A. Galgano, C.D. Blasi, S. Ritondale, A. Todisco, Numerical simulation of the glowing combustion of moist wood by means of a front-based model. Fire Mater. **38**(6), 639–658 (2013)

M. Gao, D. Pan, C. Sun, Study on the thermal degradation XE "thermal degradation" of wood treated with amino resin and amino resin modified with phosphoric acid. J. Fire Sci. **21**(3), 189–201 (2003)

S.L. Haslett, J.C. Thomas, W.T. Morgan, R. Hadden, D. Liu, J.D. Allan, P.I. Willians, S. Keita, C. Liousse, H. Coe, Highly controlled, reproducible measurements of aerosol emissions from combustion of a common African biofuel source. Atmos. Chem. Phys. **18**(1), 385–403 (2018)

M.F. Heringa, P.F. DeCarlo, R. Chirico, A. Lauber, A. Doberer, J. Good, T. Nussbaumer, A. Keller, H. Burtscher, A. Richard, B. Miljevic, A.S.H. Prevot, U. Baltensperger, Time-resolved characterization of primary emissions from residential wood combustion appliances. Environ. Sci. Technol. **46**(20), 11418–11425 (2012)

N. Ivanova, E. Bulba, Mathematical modeling of processes of heat XE "heat" and mass transfer during drying of wood biomass. Matec Web Conf. **194**, 01012 (2018)

L.S. Johansson, B. Leckner, L. Gustavsson, D. Cooper, C. Tullin, A. Potter, Emission characteristics of modern and old-type residential biomass combustion systems. Atmos. Environ. **38**(25), 4183–4195 (2004)

D. Kasymov, M. Agafontsev, Studying the effect of fire XE "fire" retardant coating on the fire hazard characteristics of wood using infrared thermography. EPJ Web Conf. **159**, 00018 (2017)

F. Kollmann, W. Coté, *Principles of Wood Science and Technology*, vol. 1 (Springer-Verlag, New York, Solid Wood, 1968)

L.A. Lowden, T.R. Hull, Flammability behaviour of wood and a review of the methods for its reduction. Fire Sci. Rev. **2**, 4 (2013)

R.E. Lyon, N. Safronava, J.G. Quintiere, S.I. Stoliarov, R.N. Walters, S. Crowley, Material properties and fire XE "fire" test XE "fire test" results. Fire Mater. **38**(2), 264–278 (2013)

R. Mensah, L. Jiang, J. Renner, Q. Xu, Characterisation of the fire XE "fire" behaviour of wood: from pyrolysis XE "pyrolysis" to fire retardant mechanisms. J. Therm. Anal. Calorim. **148**(4), 1407–1422 (2022)

V. Merk, M. Chanana, S. Gaan, I. Burgert, Mineralization of wood by calcium carbonate insertion for improved flame retardancy. Holzforschung **70**(9), 867–876 (2016)

N. Orang, H. Tran, Effect of feedstock moisture content XE "moisture content" on biomass boiler operation. Tappi J. **14**(10), 629–637 (2015)

B. Östman, E. Mikkola, L. Tsantaridis, Reaction to fire testing in Europe: the development of a European classification system. Fire Mater. **34**(9), 385–392 (2010)

B. Östman, D. Brandon, H. Frantzich, Fire safety engineering in timber buildings. Fire Saf. J. **91**, 11–20 (2017)

W.K. Purves, G.H. Orians, H.C. Heller, Life: The science of biology, Sunderland Mass., in by ed. W.H. Freeman & Co, pp. 1121 (1995). ISBN 0-7167-9856-5

Y. Qiao, H. Zhang, J. Yang, H. Chen, N. Liu, M. Xu, L. Zhang, Transition from smouldering to flaming combustion of pine needle fuel beds under natural convection. Proc. Combust. Inst. **40**(1–4), 105343 (2024)

L. Salmén, Micromechanics of the wood cell wall: a tool for a better understanding of its structure, in *Proceedings of the 1st International Conference of the European Society for Wood Mechanics, Lausanne, Switzerland*, pp. 457–470 (2001)

J. Silva, C. Castro, S. Teixeira, J. Teixeira, Evaluation of the gas emissions during the thermochemical conversion of eucalyptus woodchips. Processes **10**(11), 2413 (2022)

B.R.T. Simoneit, J.J. Schauer, C.G. Nolte, D.R. Oros, V.O. Elias, M.P. Fraser, W.F. Rogge, G.R. Cass, Levoglucosan, a biomass burning tracer for atmospheric particulate matter. Environ. Sci. Technol. **36**(4), 747–755 (2002)

A.K. Singh, R. Singh, O.P. Sinha, Characterization of charcoals produced from Acacia, Albizia and Leucaena for application in ironmaking. Fuel **320**, 123991 (2022)

Y. Sokolovskyy, I. Boretska, B.I. Gayvas, I. Kroshnyy, A. Nechepurenko, Mathematical modeling of convection drying process of wood taking into account the boundary of phase transitions. Math. Model. Comput. **8**(4), 830–841 (2021)

L.-h Tam, A. Zhou, Z. Yu, Q. Qiu, D. Lau, Understanding the effect of temperature XE "temperature" on the interfacial behavior of CFRP–wood composite via molecular dynamics simulations. Compos. Part B Eng. **109**, 227–237 (2017)

S.R. Turns, *An Introduction to Combustion: Concepts and Applications* (McGraw-Hill, New York, NY, USA, 2012)

References

B. Uysal, Combustion properties of laminated veneer lumbers bonded with polyvinyl acetate and phenol formaldehyde adhesives and impregnated with some chemicals. Combust. Sci. Technol. **177**(7), 1253–1271 (2005)

H. Wang, Y. Tian, Experimental study of the characteristic parameters of the combustion of the wood of ancient buildings. J. Fire Sci. **37**(2), 117–136 (2019)

R.H. White, Effect of lignin content and structure on the pyrolysis and combustion of wood, in *ACS Symposium Series*, vol. 376, pp. 21–34 (1987)

J.E. Winandy, R.M. Rowell, Chapter 11: chemistry of wood strength in handbook of wood chemistry and wood composites, in by ed. R.M. Rowel (CRC Press, Florida, USA, 1984), pp. 305–347

Y. Yang, T. Fu, F. Song, X. Song, X.-L. Wang, Z. Wang, Wood-burning XE "burning" processes in variable oxygen atmospheres: Thermolysis, fire XE "fire", and smoke release XE "smoke release" behavior. Polym. Degrad. Stab. **205**, 110158 (2022)

X. Zhang, W. Wang, R. Guo, W. Pan, X. Li, Study on the effect of pyrolysis conditions on the combustion behavior and char structure evolution. ACS Omega **7**, 23634–23642 (2022)

A.V. Bridgwater, Review of fast pyrolysis of biomass and product upgrading. Biomass and Bioenergy, **38**, 68–94 (2012)

B.H. Lee, H.S. Kim, S. Kim, H.J. Kim, B. Lee, Y. Deng, Q. Feng, J. Luo, Evaluating the flammability of wood-based panels and gypsum particleboard using a cone calorimeter, Constr. Build. Mater. **25**(7), 3044–3050 (2011)

J.D. McDonald, B. Zielinska, E.M. Fujita, J.C. Sagebiel, J.C. Chow, J.G. Watson, Fine particle and gaseous emission rates from residential wood combustion Environ. Sci. Technol. **34**, 2080–2091 (2000)

Chapter 3
Fire Resistance

3.1 Introduction

Timber and timber products, including laminated veneer lumber (LVL), are inherently combustible materials which raises significant concerns regarding fire safety. The fire resistance of timber in construction is a critical aspect of structural design, particularly as the use of timber in building projects continues to grow due to its sustainability and aesthetic appeal. Understanding the fire performance of timber structures is essential for ensuring the safety of occupants and the integrity of the building itself.

When timbers are exposed to flames, these materials begin to ignite at a surface temperature of approximately 270 °C. However, it is important to note that self-ignition does not occur until temperatures reach around 400 °C. According to the structural fire design guidelines outlined in Eurocode 5 (EN1995-1–2), the onset of charring is defined as the moment when the surface temperature of the timber reaches 300 °C. This charring process is crucial, as it creates a protective layer on the wood's surface that acts as an insulator, slowing down the burning process and safeguarding the underlying material from further damage.

The fire resistance behavior of wood products is highly predictable and can be calculated using the structural fire design specifications of Eurocode 5. The formation of a char layer during combustion is beneficial, as it not only insulates the remaining wood but also significantly reduces the rate of heat transfer into the material. However, the elevated temperatures that occur prior to burning can adversely affect the strength and stiffness properties of wood, even before charring begins. This reduction in mechanical properties is a critical consideration for engineers and designers when developing fire safety strategies for timber structures.

To mitigate the risk of flame spread in buildings, fire reaction requirements are established for wood surfaces. These regulations set boundary conditions for the use of visible wood in cladding and structural applications. In certain scenarios, the application of fire retardant treatments or the installation of sprinkler systems can enable the use of more visible wood elements in architectural designs, enhancing both aesthetic appeal and safety.

Furthermore, the interaction between wood and fire retardant treatments can significantly influence fire performance. Research indicates that specific chemical treatments can enhance the fire resistance of wood by altering its thermal properties and reducing its combustibility. However, it is essential to balance these treatments with potential impacts on the wood's mechanical properties, as excessive heat treatment can lead to degradation and loss of strength.

As the construction industry increasingly embraces timber as a sustainable building material, ongoing research into the fire performance of wood products is vital. Innovations in fire safety engineering and material treatments will continue to improve the viability of timber in various applications, ensuring that it can be used safely and effectively in modern construction practices. By adhering to established guidelines and incorporating advanced treatments, the construction industry can leverage the benefits of timber while maintaining the safety and resilience of structures in the face of fire hazards.

When it comes to the ability of any construction material to withstand fire, two key factors are the material's reaction to fire and its resistance to fire.

3.2 Building Fire Requirements

Fire is a powerful and destructive force that continues to pose a significant threat in Malaysia. Each year, fires result in the loss of hundreds of lives and cause extensive financial damage, amounting to millions of ringgit in property destruction and business disruption.

Concerns about fire safety and the need for protective measures to safeguard both people and property have been present in the Malaysia for centuries. Historically, early fire regulations were driven by insurance companies, with a primary focus on minimizing property damage (Bryant 2006). Over time, however, the emphasis of fire safety regulations has evolved. Today, the regulatory framework prioritizes the protection of human life (UBBL 2021), while financial losses and property damage are primarily the responsibility of insurers or building owners. This shift underscores the importance of modern fire safety measures, ensuring that buildings are designed and maintained to reduce fire risks and facilitate safe evacuation in the event of an emergency.

Safety rules for timber usage for construction are important to determine the use of various types of timber. Normally, safety rules emphasis on inhibition, detection, prevention and evacuation. Therefore, these things will definitely involve policy of design guidelines, maintenance and construction as well. Generally, building policies in Malaysia are based on the building code determined by the Malaysian Uniform Building By-Laws (UBBL). Malaysian building codes still restrict the use of timber to secondary structures rather than primary structures. This is due to a lack of data on the fire resistance of timber. In the other hand, most of the USA, Australia, and European countries have been using timber as structural application due to the durability and strength capability apart from having a unique ecstatic view.

The regulations that underpin fire safety guidance in UBBL are designed to ensure that all buildings adhere to specific functional criteria aimed at safeguarding occupants and property. These criteria encompass essential provisions for means of escape, which facilitate safe evacuation during emergencies, as well as measures to control both internal and external fire spread. Additionally, the regulations mandate the inclusion of adequate facilities for the fire service, ensuring that emergency responders can effectively manage fire incidents.

In many instances, compliance with the Building Regulations alone may not suffice to fulfill the specific safety requirements or business continuity objectives of a client. This could include advanced fire detection systems, enhanced fire suppression technologies, or specialized training for staff in emergency response procedures.

3.3 Malaysian Uniform Building by Law

Malaysian Uniform Building by Law (UBBL) requires all buildings to comply with specific functional requirements, including provisions for means of escape, control of internal and external fire spread, and adequate facilities for the fire service.

In terms of the structural fire safety functional objectives outlined in the Building Regulations for Malaysia (UBBL) 1944: Amendment 2021, several key requirements must be met to ensure the safety and integrity of buildings in the event of a fire. These objectives are critical for protecting both occupants and property, and they include the following:

- The building must be designed and constructed to maintain its structural stability for a reasonable period during a fire. This requirement ensures that the structural integrity of the building is preserved long enough for occupants to evacuate safely and for emergency services to respond effectively. The duration of stability is typically defined in relation to the building's height, occupancy type, and fire load, with specific time frames often specified in relevant fire safety codes. In general, Malaysian Building By-Laws specified that the minimum fire resistance requirement for any element of structure is between 30 min and maximum is 2 h as given in Ninth Schedule (Part VII).
- The building should be subdivided into fire-resisting compartments that are appropriately sized according to the scale and intended use of the building. This compartmentalization is essential for inhibiting the spread of fire and smoke, thereby protecting escape routes and minimizing the risk of fire affecting adjacent areas. The design of these compartments must consider various factors, including the materials used, the layout of the building, and the potential fire load within each compartment.

The interpretation and specification of provisions necessary to meet these functional requirements fall within the responsibilities of designers, particularly architects, fire engineers, and structural (fire) engineers. These professionals must work collaboratively to ensure that the design not only complies with regulatory requirements but also addresses the unique characteristics and risks associated with each building project.

Architects play a crucial role in the initial design phase, ensuring that fire safety considerations are integrated into the overall architectural vision. Fire engineers contribute their expertise in assessing fire dynamics, developing fire safety strategies, and ensuring that the building's design effectively mitigates fire risks. Structural (fire) engineers focus on the performance of materials and structural elements under fire conditions, ensuring that the building can withstand the effects of fire while maintaining its stability.

3.4 Fire Resistance

Based on Malaysian Uniform Building by Law (UBBL 1984), the requirement for fire resistance of structural member is given in Clause By-law 217.

*Any structural member or overloading wall shall have fire resistance of not less than the minimum **period** required by these By-laws for any element which it carries.*

Clause By-law 218. Compartment wall separating flat and maisonette.

Any compartment wall separating a flat or maisonette from any other part of the same building shall not be required to have fire resistance exceeding one hour unless;

(a) *the wall is a load-bearing wall or a wall forming part of a protected shaft; or*
(b) *the part of the building from which the wall separates the flat or maisonette is of a different purpose group and the minimum period of fire resistance required by this Part for any element of structure in that part is one and a half hours or more.*

The fire requirements for the materials for construction, is given in Clause By-laws 211 (UBBL 1984).

*(1) Material used in the construction of a building shall comply with the requirements stated under this part in addition to the performance requirements such as **fire resistance** or limits to **spread of flame**.*

A product's fire resistance refers to its ability to withstand fire and prevent its spread for a specified period. This characteristic plays a crucial role in fire containment, ensuring that flames, heat, and smoke are restricted to specific areas, thereby reducing the risk of fire spreading to other parts of a building. Since timber is a combustible material and can act as an additional fuel source when used in structural elements, its impact on the fire dynamics within a compartment must be carefully considered.

In the early stages of a fire, a material's reaction to fire properties is of primary importance. These properties determine how quickly the material ignites, how much

3.4 Fire Resistance

heat and smoke it generates, and whether it contributes to flame propagation. Materials with poor reaction-to-fire performance can accelerate the growth of a fire, making early intervention more difficult.

However, once the fire has fully developed, fire resistance becomes the critical factor. At this stage, structural and compartmentalizing elements, such as fire doors, walls, floors, and ceilings, must function effectively to contain the fire within designated compartments. This containment helps maintain structural integrity, provides safe escape routes for occupants, and allows firefighters sufficient time to control and extinguish the blaze.

Generally, fire performance on timber will go through three stages as shown in Fig. 3.1 and summarized as follows:

a. Growth Phase: This initial stage describes how the material reacts to fire which begins with the ignition of the timber, where small flames start to develop. The fire gradually increases in intensity as it consumes available combustible materials. During this phase, the temperature rises, and the fire spreads to adjacent surfaces, leading to a build-up of heat and smoke. The duration of this phase can vary significantly based on factors such as the type of timber, moisture content, and the presence of fire-retardant treatments.
b. Post-flashover Phase: Once the fire reaches flashover, it transitions into the fully developed fire stage. Flashover is a critical moment when the temperature within the space becomes high enough to ignite all combustible materials simultaneously. This stage is marked by rapid fire spread and a significant increase in heat release rate, resulting in a dramatic rise in temperature. In this phase, the fire can

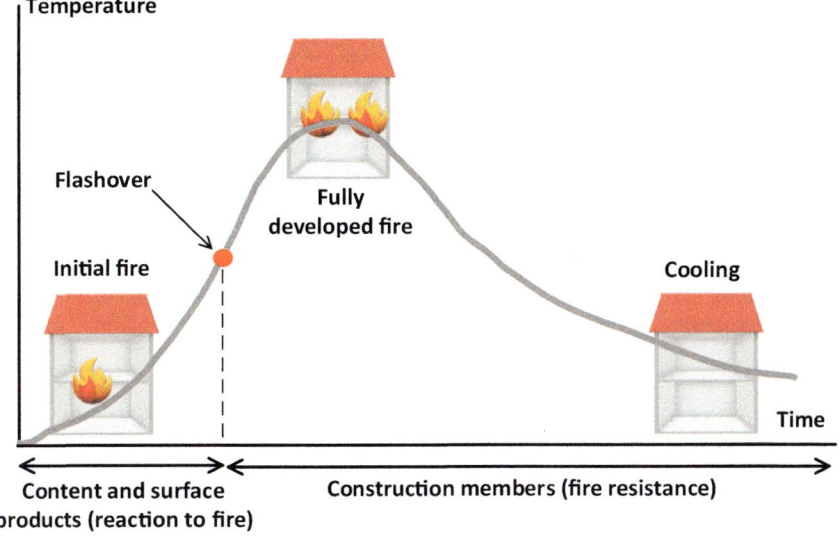

Fig. 3.1 Two main stages relevant for the fire safety in buildings in relation to building materials and structures

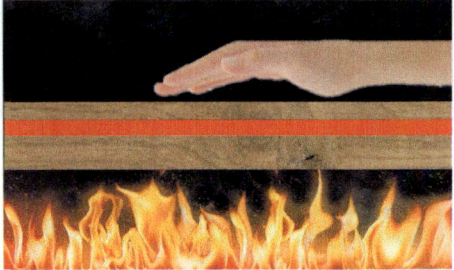

Integrity E
No flames passing to unexposed side during the certified period.

Insulation I
Temperature shall never increase more than 140°C on average at unexposed side during the certified period.

Fig. 3.2 Schematic diagram on the resistance to fire

reach its peak intensity, posing severe risks to structural integrity and occupant safety.

c. Decay Phase: Following the fully developed fire, the decay phase begins as the fuel sources are depleted. During this stage, the temperature gradually declines as the fire consumes the remaining combustible materials. The intensity of the flames diminishes, and the smoke production decreases. This phase is crucial for evacuation efforts, as it may provide a window of opportunity for occupants to escape if they have not already done so.

Understanding these stages is essential for effective fire safety design and management, particularly in timber structures. By recognizing the behavior of timber during each phase of a fire, architects, engineers, and safety professionals can implement appropriate fire protection measures and design strategies to enhance the resilience of timber buildings against fire hazards.

The resistance to fire of a product are indicated as a time duration. It assesses if the product can resist fire and prevent it from spreading to the opposing side for 30, 60, 90, 120, or 240 min and are determined through standardized fire tests. These tests evaluate the ability of building components to withstand fire exposure and are assessed based on three key performance criteria as illustrated in Fig. 3.2.

a. Load-Bearing Capacity (R): The ability of a structural element to maintain its stability under fire conditions.
b. Integrity (E): The ability to prevent the passage of flames and hot gases.
c. Insulation (I): The ability to limit the temperature rise on the unexposed side of the element.

Fire resistance ratings are determined through standardized fire tests, such as those outlined in BS 476 or EN 1363 or ISO 834, where building materials and components are subjected to controlled fire conditions to evaluate their performance. These ratings help ensure that a building's structure and fire compartments provide sufficient time for occupants to evacuate and for emergency services to respond effectively.

The higher the fire resistance rating, the longer the product can endure fire conditions, contributing to overall building safety and fire management.

3.5 Reaction to Fire

Reaction to fire assesses how materials behave when exposed to fire or heat, considering factors such as ignitability, flame spread, heat release, smoke production, and the potential to generate flaming droplets. Reaction to fire testing primarily evaluates a construction material's contribution to fire development, particularly during the initial stages of ignition and fire growth.

When exposed to high temperatures, wood undergoes thermal degradation (pyrolysis), leading to physical, structural, and chemical changes. Heat is primarily transferred to the surface of timber through radiation and convection, after which it is conducted deeper into the material. The char layer acts as an insulation layer, which slows the burning and protects the rest of the cross-section (see Charred samples for Light Red Meranti after 30 min after one-dimensional fire exposure Figs. 3.3 and 3.4). The key factors influencing this process are exposure time and temperature intensity.

The fire performance of timber is influenced by various temperature regimes, which lead to significant changes in its mechanical and thermo-physical properties. Due to the thermally thick nature of timber, a temperature gradient develops within the material when exposed to heat.

Initially, heated timber begins to lose its strength at temperatures as low as 65 °C Palma et al. (2016). As the temperature rises to around 100 °C, moisture within the wood starts to evaporate, resulting in a plateau in the temperature distribution known as the latent heat of vaporization (Bartlett et al. 2019). This phenomenon indicates

Fig. 3.3 Charred samples for light red meranti after 30 min after one-dimensional fire exposure

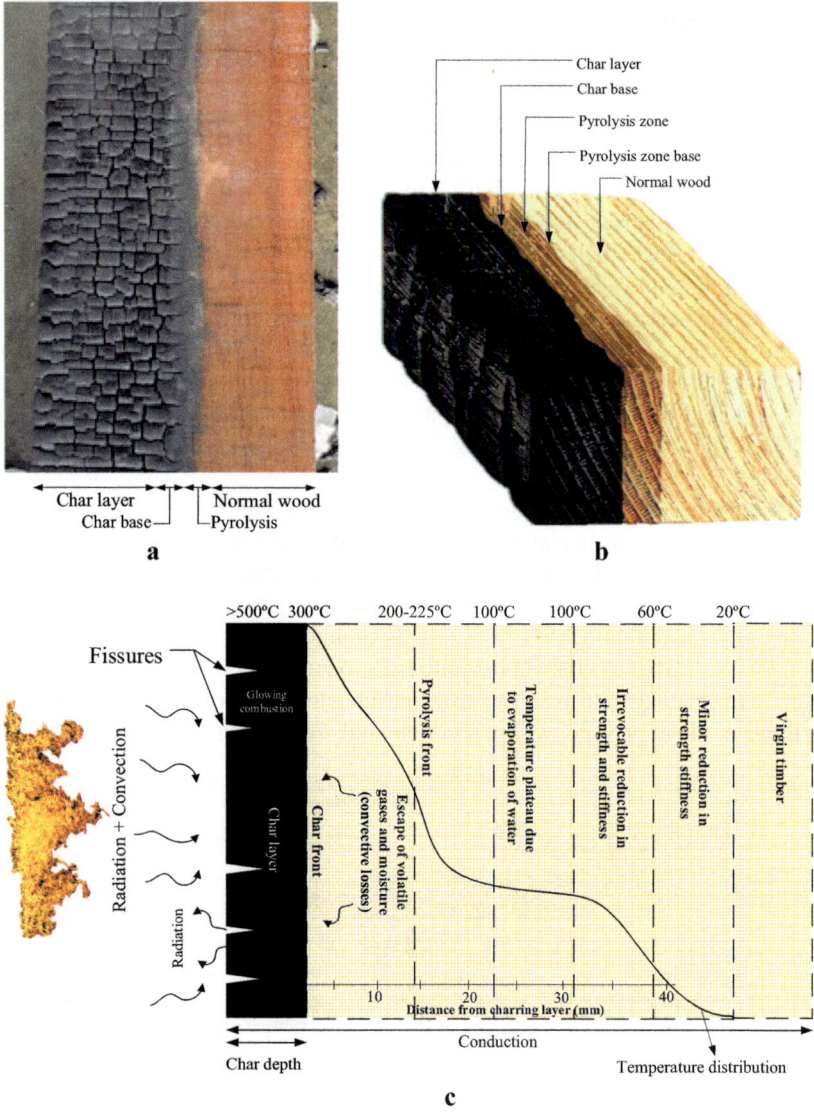

Fig. 3.4 Cross-section of fire-exposed timber; **a** actual charred sample, **b** illustration of charring of wood beam, **c** charring of wood beam with temperature gradient in burning wood, illustrating the heat transfer process and internal thermal degradation

3.5 Reaction to Fire

that energy is being used to convert water from liquid to vapor rather than increasing the temperature of the wood itself. The onset of pyrolysis, a chemical decomposition process that occurs when timber is subjected to high temperatures, typically begins at temperatures ranging from 200 to 225 °C (Žajdlík and Šuhajda 2022). Continued exposure to heat results in an increase in the depth of this charred layer, which further contributes to mass loss and reduces the effective cross-section of the timber.

- The char layer that forms during a fire is beneficial as it acts as an insulating barrier, protecting the unburned wood beneath from direct heating. This layer also inhibits the escape of volatile gases, thereby reducing the thermal penetration depth and the rate of pyrolysis (Moss et al. 2009; Létourneau-Gagnon et al. 2021). However, this protective effect can be compromised by surface regression, or shrinkage, of the char layer, which occurs at temperatures exceeding 550 °C due to the oxidation of char. This regression results in a reduction of char depth, consequently diminishing the protection afforded to the underlying wood (Xu et al. 2016; König 2006).
- Moreover, the development of fissures or cracks within the char layer can further hinder its protective capabilities, allowing for increased heat transfer into the timber. This phenomenon underscores the importance of understanding the thermal behavior of timber in fire scenarios, as it directly impacts the structural integrity and safety of timber constructions during fire events.
- At around 300 °C, wood begins to degrade rapidly, forming a carbonaceous char layer. This layer develops with cracks and fissures, allowing heat to penetrate deeper into the timber section. However, from a fire resistance perspective, the formation of the char layer is beneficial, as it acts as an insulating barrier, protecting the underlying, unaffected wood. Burning creates a char layer on the surface of wood products.

Exposure to elevated temperatures prior to the onset of charring significantly diminishes the strength and stiffness of wood. This reduction in mechanical properties is critical and must be carefully considered in structural fire design to ensure that timber performs adequately under fire conditions as illustrated in Fig. 3.5.

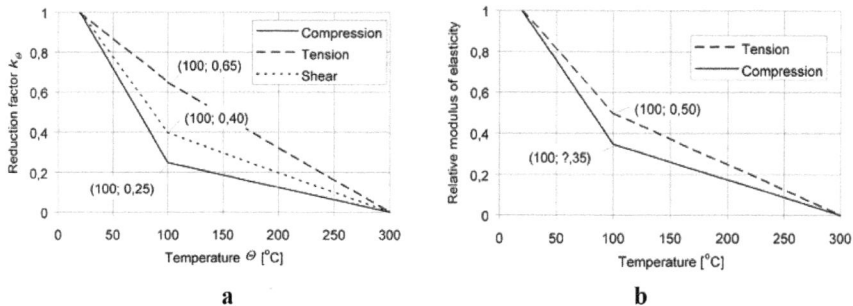

Fig. 3.5 Effect of temperature on the mechanical properties of softwood; **a** reduction factor for strength parallel to grain, **b** reduction factor for modulus of elasticity parallel to grain (EN1995-1-2:2004, Fig. B.4 and B.5, respectively)

Fig. 3.6 Reaction to fire test for glulam from Malaysian tropical timber using room corner test

Reaction to fire requirements is established for wood surfaces to mitigate the risk of flame spread within buildings. These regulations set specific boundary conditions that govern the use of visible wood in cladding and structural applications, ensuring that fire safety is maintained while allowing for aesthetic considerations in architectural design.

The primary goal of these requirements is to prevent the rapid spread of flames across surfaces, which can pose significant hazards to occupants and property. By implementing these regulations, building codes aim to create safer environments that minimize the potential for fire-related incidents. The requirements often dictate the fire performance classification of building products, which is determined through standardized testing methods. The Reaction to Fire classification system was established using a large-scale Room Corner Test, often referred to as a "reference test". Figure 3.6 shows the reaction to fire test according to EN 13,823.

The classifications are determined based on the likelihood of commonly used construction materials reaching flashover during the test. These classifications help architects and builders select appropriate materials that meet safety standards while still achieving the desired visual appeal.

3.6 Reaction to Fire Classification

The reaction to fire performance classes for different construction material applications are given in the European classification system defined in EN 13,501–1, as shown in Table 3.1. In the fire classification system, the first letter (A–E) indicates combustibility, with D being the typical rating for wood products. The second symbol (s1–s3) represents smoke production, while the third symbol (d0–d3) denotes the risk

3.6 Reaction to Fire Classification

of flaming droplets. Table 3.2 further explains the performance criteria for reaction to fire.

Laminated veneer lumber (LVL) generally falls within the same classification as most untreated wood products. In the European classification system (EN 13,501–1), LVL is classified as D-s2, d0 for its reaction to fire performance. This classification applies to LVL without the need for additional testing, provided that the density is at least 400 kg/m^3 and the thickness is ≥ 18 mm, as specified in Table 3.1 of the European Commission Delegated Regulation (EU) 2017/2293 and reproduced here as Table 3.3.

A study was conducted to evaluate the reaction to fire of glulam product made from Malaysian tropical timber species, specifically Malagangai in accordance with EN 13,501–1: standard for fire classification of construction materials using single burning item test (EN 13,823). In determining the flame spread, the measurement of average heat release-rate HRRav(t), the total heat release THR(t) and the FIGRA(t)-value for Malagangai glulam as shown in Fig. 3.7a and b respectively were analyzed. As for the classification of the smoke production, the graphs in Fig. 3.7c and d were analyzed.

Table 3.1 Overview of the European reaction to fire classes for building products excluding floorings (EN 13,501–1)

Euroclass	Smoke class	Burning droplets class	Requirements according to			FIGRA W/s	Typical products
			Non comb	SBI	Small flame		
A1	–	–	x	–	–	–	Stone, concrete
A2	s1, s2 or s3	d0, d1, or d2	x	x	–	≤120	Gypsum board (thin paper), mineral wool
B	s1, s2 or s3	d0, d1, or d2	–	x	x	≤120	Gypsum boards (thick paper), fire retardant (wood)
C	s1, s2 or s3	d0, d1, or d2	–	x	x	≤250	Coverings on gypsum boards
D	s1, s2 or s3	d0, d1, or d2	–	x	x	≤750	Wood, wood-based panels
E	–	- or d2	–	–	x	–	Some synthetic polymers
F	–	–	–	–	–	–	No performance determined

SBI = Single Burning Test (EN 13,823), main test for the reaction to fire classes for building products;
FIAGRA—Fire Growth Rate, main parameter for the main fire class according to the SBI test.

Table 3.2 Performance criteria for reaction to fire test for building products

Class	Reaction to fire	Flashover in the room corner reference test	Additional criteria tested for
A1	No contribution to a fire	No	None (insignificant smoke release with no flaming droplets or particles expected)
A2	No significant contribution to fire growth	No	Production of smoke & flaming droplets or particles
B	Very limited contribution to fire growth	No	Production of smoke & flaming droplets or particles
C	Limited contribution to flashover	Flashover after 10 min	Production of smoke & flaming droplets or particles
D	Contribution to flashover	Flashover between 2 to 10 min	Production of smoke & flaming droplets or particles
E	Significant contribution to flashover	Flashover before 2 min	Production of flaming droplets or particles (Smoke release is expected to be substantial)
F	Not tested or incapable of achieving Class E	No performance determined	

Table 3.3 Classes of reaction to fire performance for cross laminated timber products and laminated veneer lumber products for walls and ceilings

Product ([1])	Product detail	Minimum mean density ([2]) (kg/m^3)	Minimum overall thickness (mm)	Class ([3])
Cross laminated timber products covered by the harmonized standard EN 16,351	Minimum layer thickness of 18 mm	350	54	D-s2, d0([4])
Laminated veneer lumber products covered by the harmonized standard EN 14,374	Minimum veneer thickness of 3 mm	400	18	D-s2, d0([4])

([1]) Applies to all species and glues covered by the product standards.
([2]) Conditioned in accordance with standard EN 13,238.
([3]) Class as set out in Table 3.1 of the Annex to Delegated Regulation (EU) 2016/364.
([4]) Class valid for any substrate or air gap behind.

3.6 Reaction to Fire Classification

The results, presented in Fig. 3.7, indicate that Malagangai glulam falls under Class C with fire growth rate (FIGRA) ≤ 250 Ws-1, demonstrating moderate resistance to flame spread with minimal smoke production and limited formation of burning droplet as seen in Fig. 3.7e.

In Malaysia, the classification of building materials based on surface flame spread characteristics follows the British Standard BS 476: Part 7. This standard categorizes

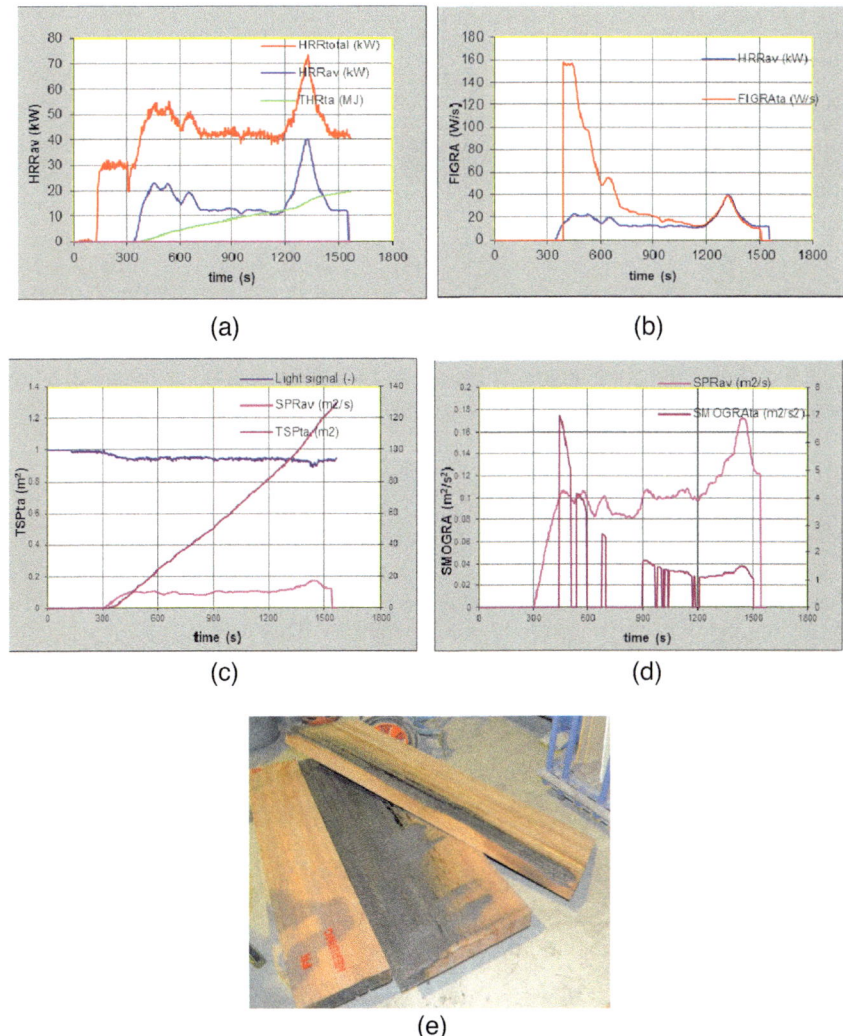

Fig. 3.7 Graphs of reaction to fire; **a** the average heat release-rate HRRav(t) and the total heat release THR(t), **b** the FIGRA(t)-value, **c** the smoke production-rate SPRav(t) and the total smoke production TSP(t), **d** the SMOGRA(t)-value, **e** samples after reaction to fire test

Table 3.4 Comparison of European and UK fire standards/classifications

European		UK	
Standard	Classification	Standard	Classification
ISO 1182 and ISO 1716	Class A1	BS 476: Part 4	Non combustible
ISO 1182 or ISO 1716 and EN 13,823	Class A2	BS 476: Part 11	Limited combustibility
EN 13,823 and EN ISO 11925–2	Class B	BS 476: Parts 6 and 7	*Class 0
EN 13,823 and EN ISO 11925–2	Class C	BS 476: Part 7	*Classes 1 and 2
EN 13,823 and EN ISO 11925–2	Class D	BS 476: Part 7	*Class 3
EN 13,823 and EN ISO 11925–2	Class E	BS 476: Part 7	*Class 4

Note * Malaysia flame spread classes

materials into Classes 0, 1–4, with Class 1 indicating the lowest rate of flame spread and Class 4 the highest. Malagangai glulam falls into Class 2 for flame spread. Table 3.4 shows the comparison of European and UK standards and classification for fire reaction in testing of materials and products.

3.7 Charring of Wood

Charring is the process in which a char layer forms on the burning surface of a timber member when exposed to high temperatures. This process begins when the wood reaches approximately 300 °C, leading to thermal degradation (pyrolysis), which results in the formation of three distinct layers:

a. Char Layer—The outermost burnt layer, which acts as an insulator.
b. Pyrolysis Zone—The intermediate layer where decomposition occurs.
c. Residual Timber—The unaffected inner core that retains its original structural properties.

These degradation zones are illustrated in Fig. 3.4b, highlighting the progression of charring under fire conditions (White 2008). Charring behavior in timber can be categorized into two primary types:

a. One-Dimensional Charring

A fundamental property of a specific wood species, or timber classified by density and strength. The charring rate in this category is linear, meaning the char depth increases steadily over time under standard fire exposure. It is commonly referred to as the one-dimensional charring rate, which is particularly relevant for unprotected timber slabs where heat transfer occurs in a single direction.

b. Two-Dimensional Charring

This charring takes into accounts for cross-sectional effects, where fire exposure occurs from multiple directions. The charring rate can vary depending on timber size, grain orientation, and fire dynamics. It is more complex than one-dimensional charring, as it influences the structural integrity of the timber differently.

As the fire continues, the temperature within the timber increases, causing the boundary between the unaffected wood and the pyrolysis zone to gradually shift deeper into the material. The rate at which the boundary between burnt and unburnt wood progresses through the cross-section of the material when exposed to fire is referred to as the charring rate of timber. The charring rate is determined from the distance between the outer surface of the original member and the position of the char-line as illustrated in Fig. 3.8 over the time of exposure (Cachim and Franssen 2009). It is measured in millimeters per minute (mm/min) and serves as a critical parameter for assessing fire resistance in timber structures.

The charring rate depends on various factors, such as timber species and density (denser woods tend to char more slowly), moisture content (higher moisture delays charring due to water evaporation), fire exposure conditions (including temperature, duration, and oxygen availability), and the presence of fire retardants or surface treatments.

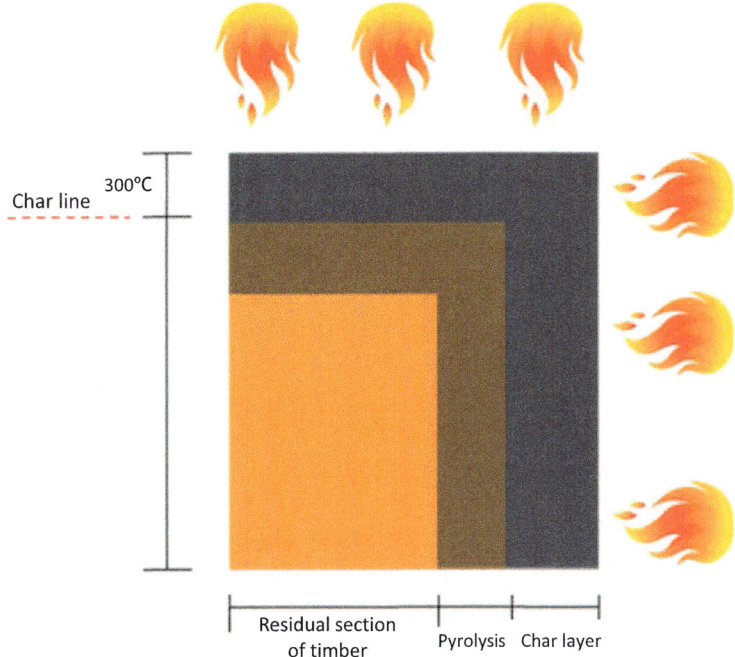

Fig. 3.8 Changes of region of timber when subjected to fire (Cachim and Franssen 2009)

Furthermore, the charring rate plays a crucial role in fire design, as it dictates how rapidly the load-bearing section of timber decreases to a critical level. However, the charred layer lacks rigidity and strength, rendering it incapable of supporting any loads. Determining the charring rate of timber is essential in structural fire design to assess the fire resistance and load-bearing capacity of timber elements when exposed to high temperatures (Lau et al. 1999).

Eurocode 5 (EC5) permits both simplified and advanced methods for calculating the extent of charred timber. The simplified approach utilizes a charring rate model, which, despite the nonlinear progression of charring depth over time, assumes a linear behavior due to its limited nonlinearity in practical applications. This allows the charring rate to be considered constant over time. The strength of the remaining cross-section is then assessed under the assumption that its mechanical properties remain comparable to those at room temperature.

3.7.1 One-Dimensional Charring Rate

The one-dimensional charring rate refers to the rate at which charring progresses in a single direction due to heat transfer under standard fire exposure conditions in an unprotected, semi-infinite timber slab without any fissures or gaps (FireInTimber Project 2010) as shown in Fig. 3.9. This rate, often referred to as the linear charring rate, remains relatively constant after an initially higher charring phase.

The charring depth refers to the distance from the original outer surface of a timber member to the char line. The charring depth, d_c, is the absolute decrease of the timber cross-section at a point in time due to a fire as illustrated in Fig. 3.10. Since charring depth, d_c is determined as a function of time, the charring rate, β, can be derived using Eq. 3.1.

Fig. 3.9 Illustration on the one-dimensional fire exposure

3.7 Charring of Wood

Fig. 3.10 One-dimensional charring of wide cross-section (fire exposure for one side)

The charring rate for one-dimensional fire exposure can be calculated as:

$$\beta_t = \frac{d_c}{t}, \qquad (3.1)$$

where: β_t = charring rate (mm/min).
d_c = charring depth (mm).
t = duration of fire (min).

In design, the one-dimensional charring rate is denoted as $d_{char,0}$. The charring rate of unprotected building members in standard fire exposure under one-dimensional heat transfer is usually assumed as the base value in the design of timber buildings and structures (Aseeva et al. 2014).

3.7.2 Two-Dimensional Charring Rate

The application of the two-dimensional charring rate is particularly relevant in beam or column structures and similar architectural elements such in Fig. 3.11. It is used to assess how timber beams or members behave when exposed to fire from multiple sides, allowing engineers to design these elements with a greater understanding of their fire performance and safety characteristics.

For two-dimensional charring rate, it basically involves the effects of cross-sectional dimensions. When a rectangular timber section is subjected to exposure

Fig. 3.11 Beam with potential for three-sided fire exposure conditions

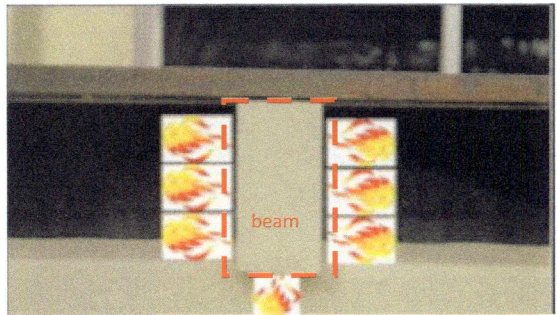

to fire, the corners of the timber experience heat transfer from two adjacent surfaces, leading to a phenomenon known as corner rounding (Buchanan 2002). Initially, this rounding at the arris where two edges of the timber meet, is approximately equivalent to the depth of charring observed in a one-dimensional context. However, as the fire exposure continues, a significant interaction occurs between the rounding effects of the two opposing arrises. This interaction results in an increased charring depth on the narrower side of the rectangular cross-section compared to the wider side.

For understand ability, notional charring rates were introduced in order to transform the unequal shape of residual cross-sections into simple rectangular cross-sections (Frangi and König 2011) and EC5 has simplified the residual cross-section by replacing the one-dimensional charring depth and arris rounding with an equivalent notional charring depth (see Figs. 3.12 and 3.13).

Two-dimensional or notional charring rate, β_n, is calculated as:

$$\beta_n = \frac{d_c}{t}. \tag{3.2}$$

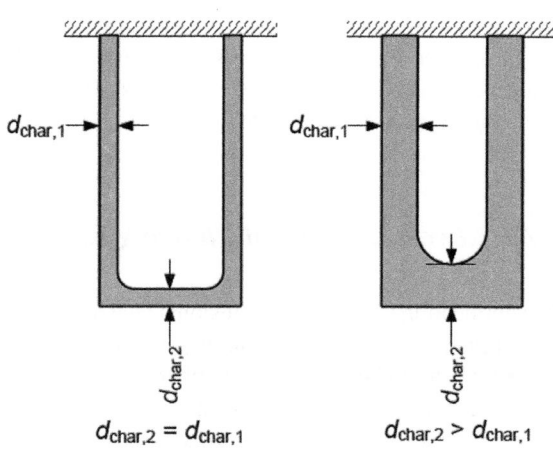

Fig. 3.12 Effect of arris rounding on charring on the wide and narrow sides of cross-section and charring depth measurement (FireInTimber Project 2010)

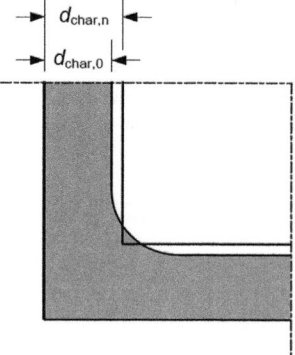

Fig. 3.13 Charring depth for one-dimensional charring and notional charring depth (FireInTimber Project 2010)

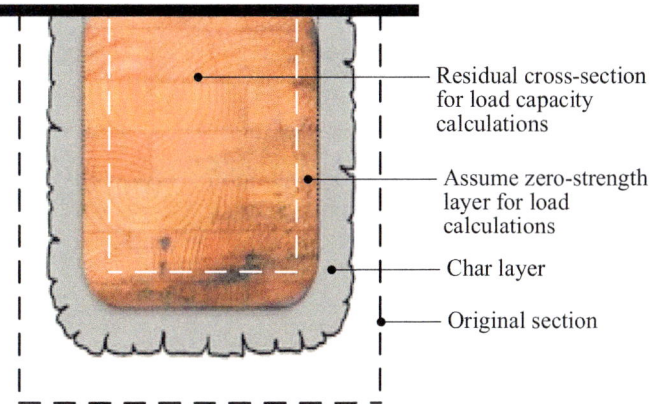

Fig. 3.14 Graphic process of charring rate model of timber beam (White 2013)

Normally, the amount of the charring depth is gained either during the fire test or after the fire test, which is usually done continuously basis (Tsai 2010). Figure 3.14 illustrates the standard process of notional charring rate of timber when exposed to three-sided fire exposure.

3.8 Fire Resistance Considerations in Timber Structure Design: The Malaysian Context

The requirement for fire safety can be obtained by full-scale testing of building construction or by design of building construction by means of calculations according to design standards which is by using the charring rate value.

Currently, in Malaysia, the structural design of timber is still based on the permissible stress design as given in a series of Malaysian Standards for timber structures; MS 544. The MS 544 is design code for timber structures which contains 12 parts as follows:

Part 1: General.
Part 2: Permissible stress design of solid timber.
Part 3: Permissible stress design of glued laminated timber.
Part 4: Timber panel products:

   Section 1: Structural plywood.
   Section 2: Marine plywood.
   Section 3: Cement bonded particle board.
   Section 4: Oriented strand board.

Part 5: Timber joints.
Part 6: Workmanship, inspection and maintenance.

Part 7: Testing.
Part 8: Design, fabrication and installation of prefabricated timber for roof trusses.
Part 9: Fire resistance of timber structures.

Section 1: Method of calculating fire resistance of timber members.

Part 10: Preservative treatment of structural timbers.
Part 11: Recommendation for the calculation basis for span tables.

Section 1: Domestic floor joists.
Section 2: Ceiling joists.
Section 3: Ceiling binders.
Section 4: Domestic rafters.

Part 12: Laminated veneer lumber for structural application.

In Malaysia, the design of solid sawn timber structures adheres to the guidelines specified in Malaysian Standard MS 544: Part 2: 2017. This standard adopts BS 5268-2: 1991 in terms of design procedures but not the strength data. The strength data in BS 5268 are based on large size specimens which cover softwood and hardwoods from Europe as well as some timbers from other countries such as Australia, New Zealand, America, Canada, South Africa, and a few other Southeast Asia countries.

The strength data in Table 3.4 of MS 544: Part 2 are tabulated here as Table 3.5 which were derived from small clear specimen. These strength data are converted into grade stresses and then grouped together with similar grade stresses and density.

Therefore, when designing fire-resistant timber structures according to MS 544: Part 9, the classification of charring rates is also based on strength groups (SG1 to SG7), as detailed in Table 3.6. This approach differs from BS 5268 (Table 3.7) and Eurocode 5 (Table 3.8), which categorize charring rates based on timber density rather than strength groups.

In Eurocode 5, charring rates are classified into two distinct categories: one-dimensional and two-dimensional charring rates. At present, Malaysian Standard MS 544: Part 9 provides charring rates for solid timber but does not include provisions for engineered timber products.

These variations highlight the differences in fire resistance approaches between the Malaysian and British standards, reflecting the need to consider regional timber characteristics and fire safety requirements in structural design.

In order to move toward limit state design, the charring rate in MS 544 Part 9 has to be changed. This book incorporates the charring rate for laminated veneer lumber from Malaysian tropical timber in accordance with Eurocode 5.

3.8 Fire Resistance Considerations in Timber Structure Design: The … 59

Table 3.5 Wet and dry grade stress for various strength groups of timber (stresses and moduli expressed in N/mm² (*Source* Table 14, MS 544 Part 2)

Strength groups	Condition[1]	Bending parallel to grain			Tension parallel to grain			Compression parallel to grain			Compression perpendicular to grain[2]				Shear parallel to grain			Modulus elasticity for all grades	
		Sel	Std	Com[3]	Sel	Std	Com	Sel	Std	Com	Basic	Sel	Std	Com	Sel	Std	Com	Mean	Minimum
SG1	Wet	29.2	23.0	18.2	17.5	13.8	10.9	26.8	21.1	16.8	4.59	3.90	3.67	3.44	2.54	1.98	1.59	17,000	13,300
	Dry	33.6	26.5	21.0	20.2	15.9	12.6	28.5	22.5	17.8	4.67	3.97	3.74	3.50	2.94	2.28	1.64	18,800	14,000
SG2	Wet	20.7	16.3	13.0	12.4	9.8	7.8	18.8	14.8	11.7	3.50	2.97	2.80	2.62	2.24	1.74	1.40	15,700	11,700
	Dry	23.3	18.3	14.6	14.0	11.0	8.8	23.4	18.5	14.7	3.82	3.25	3.05	2.86	2.51	1.95	1.57	16,800	12,600
SG3	Wet	18.1	14.2	11.3	10.9	8.5	6.8	15.3	12.0	9.5	2.38	2.02	1.90	1.78	1.84	1.43	1.15	13,300	9800
	Dry	20.2	15.9	12.6	12.1	9.5	7.6	17.8	14.1	11.1	2.61	2.22	2.09	1.96	2.07	1.61	1.30	14,300	10,300
SG4	Wet	14.2	11.2	8.8	8.5	6.7	5.3	12.1	9.5	7.6	1.83	1.55	1.46	1.37	1.53	1.19	0.96	10,700	7400
	Dry	16.8	13.2	10.5	10.1	7.9	6.3	14.1	11.1	8.8	2.06	1.75	1.65	1.54	1.58	1.23	0.99	11,000	7600
SG5	Wet	11.0	8.6	6.8	6.6	5.2	4.1	9.1	7.2	5.7	1.12	0.95	0.90	0.84	1.21	0.95	0.76	8800	6100
	Dry	12.1	9.5	7.5	7.3	5.7	4.5	10.8	8.5	6.7	1.42	1.21	1.14	1.06	1.37	1.07	0.86	9100	6300
SG6	Wet	9.4	7.4	5.9	5.6	4.4	3.5	7.9	6.2	5.0	1.02	0.87	0.82	0.76	1.05	0.82	0.66	6700	4900
	Dry	11.3	8.9	7.1	6.8	5.3	4.3	8.8	6.9	5.5	1.28	1.09	1.02	0.96	1.11	0.86	0.69	7300	5200
SG7	Wet	6.6	5.2	4.2	4.0	3.1	2.5	5.3	4.2	3.3	0.62	0.53	0.50	0.46	0.91	0.71	0.57	5700	3000
	Dry	8.2	6.5	5.1	4.9	3.9	3.1	6.9	5.4	4.3	0.77	0.65	0.62	0.58	0.98	0.76	0.61	6600	3400

Table 3.6 Charring rate from MS 544: Part 9

Species	Charring rate (mm/min)
SG 1 to SG3	0.5
SG 4 to SG 5	0.7

Table 3.7 Charring rate from BS5268: Part 4

Species	Charring rate (mm/min)
Softwood <640kg/m^3	0.83
Hardwood >640kg/m^3	0.55
Softwood glued laminated timber	0.70

Table 3.8 Design charring rates and of timber

Species	Charring rate (mm/min)	
	β_0	β_n
(a) Softwood and beech Glued laminated timber with a characteristic density of ≥ 290 kg/m^3 Solid timber with a characteristic density of ≥ 290 kg/m^3	0,65 0,65	0,7 0,8
(b) Hardwood Solid or glued laminated hardwood with a characteristic density of 290 kg/m^3 Solid or glued laminated hardwood with a characteristic density of ≥ 450 kg/m^3	0,65 0,50	0,7 0,55
(c) LVL With a characteristic density of ≥ 480 kg/m^3	0,65	0,7
(d) Panels Plywood	0,9[a]	–
Wood panelling	0,9[a]	-
Wood-based panel other than plywood	1,0[a]	–

[a]The values apply to characteristic density of 450 kg/m^3 and a panel thickness of 200 mm; 3.4.2 (9) for other thicknesses and densities.
Source EC5.

3.9 Summary

In conclusion, the fire resistance of timber structures is a multi-faceted issue that requires a comprehensive understanding of material properties, connection performance, protective measures, and modeling techniques. Continued research in these areas is essential for advancing the safety and performance of timber in construction, particularly as the demand for sustainable building materials grows.

References

A.I. Bartlett, R.M. Hadden, L.A. Bisby, A review of factors affecting the burning behaviour of wood for application to tall timber construction. Fire Technol. **55**(1), 1–49 (2019)

BS 476–22:1987—Fire tests on building materials and structures—Part 22: Methods for determination of the fire resistance of non-loadbearing elements of construction. BSI Standards Limited

A. Buchanan, *Structural Design for Fire Safety* (John Wiley & Sons, Chichester, UK, 2002)

A. Buchanan, B. Östman, *Fire Safe Use of Wood in Buildings* (2022).https://doi.org/10.1201/978 1003190318

P.B. Cachim, J. Franssen, Comparison between the charring XE "charring" rate model and the conductive model of Eurocode 5. Fire Mater. **33**(3), 129–143 (2009)

Commission Delegated Regulation (EU) 2017/2293 of 3 August 2017. on the conditions for classification, without testing, of laminated veneer lumber products covered by the harmonised standard EN 14374 with regard to their reaction to fire

BS EN 1363–1:2020—Fire resistance tests—Part 1: General requirements. BSI Standards Limited

BS EN 13381–7:2019—Test methods for determining the contribution to the fire resistance of structural members—Part 7: Applied protection to timber members. BSI Standards Limited

C. Erchinger, A. Frangi, M. Fontana, Fire design of steel-to-timber dowelled connections. Eng. Struct. **32**(2), 580–589 (2010)

W.K. Tsai, *Charring Rates for Different Cross Sections of Laminated Veneer Lumber (LVL)*. University of Canterbury (2010)

A. Frangi, König, J., Effect of increased charring on the narrow side of rectangular timber cross-sections exposed to fire on three or four sides (January), 593–605 (2011)

J. König, Effective thermal actions and thermal properties of timber members in natural fires. Fire Mater. an Int. J. **30**(1), 51–63 (2006)

D. Lange, L. Boström, J. Schmid, J. Albrektsson, The reduced cross section method applied to glulam timber exposed to non-standard fire XE "fire" curves. Fire Technol. **51**(6), 1311–1340 (2015)

J. Laranjeira, H. Cruz, A. Pinto, C. Santos, J. Pereira, Reaction to fire XE "fire" of existing timber elements protected with fire retardant treatments: experimental assessment. Int. J. Arch. Herit. **9**(7), 866–882 (2014)

P.W.C. Lau, R. White, Zeeland, Van I, Modelling the charring behaviour of structural lumber, **216**, 209–216 (1999)

M. Létourneau-Gagnon, C. Dagenais, P. Blanchet, Fire performance of self-tapping screws in tall mass-timber buildings. Appl. Sci. **11**(8), 3579 (2021)

R. Mensah, L. Jiang, J. Renner, Q. Xu, Characterisation of the fire XE "fire" behaviour of wood: from pyrolysis XE "pyrolysis" to fire retardant mechanisms. J. Therm. Anal. Calorim. **148**(4), 1407–1422 (2022)

P. Moss, A. Buchanan, M. Fragiacomo, C. Austruy, Experimental testing and analytical prediction of the behaviour of timber bolted connections subjected to fire XE "fire." Fire Technol. **46**(1), 129–148 (2009)

MS 544–2:2001—Code of Practice for Structural Use of Timber – Part 2: Permissible Stress Design of Solid Timber. (2001). Department of Standards Malaysia.

MS 544–9–1:2001—Code of practice for structural use of timber—Part 9: Fire Resistance of Timber Structures—Section 1: Method of Calculating Fire Resistance of Timber Structures. (2001). Department of Standards Malaysia

A. Mydin, N. Sani, N. Abas, Y. Khaw, Evaluation of fire XE "fire" hazard and safety management of heritage buildings in georgetown, penang. Matec Web of Conferences **10**, 06003 (2014)

B. Östman, L. Tsantaridis, Fire scenarios for multi-storey façades with emphasis on full-scale testing of wooden façades. Fire Technol. **51**(6), 1495–1510 (2015)

P. Palma, A. Frangi, E. Hugi, P. Cachim, H. Cruz, Fire resistance XE "Fire resistance" tests on timber beam-to-column shear connections. J. Struct. Fire Eng. **7**(1), 41–57 (2016)

Uniform Building By-Laws 1984 (Amendment 2021). Government of Malaysia

R.H. White, Analytical methods for determining fire resistance of timber members, Section 4, Chapter 13, SFPE Handbook of Fire Protection Engineering (4th Edition), Society of Fire Protection Engineers (2008)

F. Wiesner, A. Bartlett, S. Mohaine, F. Robert, R. McNamee, J. Mindeguia, L. Bisby, Structural capacity of one-way spanning large-scale cross-laminated timber slabs in standard and natural fires. Fire Technol. **57**(1), 291–311 (2020)

Q. Xu, W. Yong, L. Chen, R. Gao, X. Li, Comparative experimental study of fire XE "fire" -resistance ratings of timber assemblies with different fire protection measures. Adv. Struct. Eng. **19**(3), 500–512 (2016)

T. Žajdlík, K. Šuhajda, Experimental testing results of a timber structure's fire XE "fire" resistance in a combustion chamber. Key Eng. Mater. **932**, 225–230 (2022)

R. Aseeva, B. Serkov, A. Sivenkov, Fire Behavior and Fire Protection in Timber Buildings (Springer Series in Wood Science), R. Wimmer (Ed.), Vol. IX, (Springer Dordrecht Heidelberg New York London, 2014)

FireInTimber Project, Fire safety in timber buildings: Technical guideline for Europe (SP Report 2010:19). SP Technical Research Institute of Sweden, (2010)

Chapter 4
One-Dimensional Charring Rate for Laminated Veneer Lumber from Malaysian Tropical Timber

4.1 Introduction

The fire resistance of wooden structural elements is heavily influenced by the rate at which wood burns. To address this, various building codes, such as EN 1995-1-2, MS 544: Part 9 and AS 1720.4 provide standardized models for calculating the amount of timber that chars over a given period during a fire event. These models are crucial for assessing the structural integrity of timber elements under fire exposure.

When exposed to fire, timber undergoes a degradation process, forming a charred layer that separates the unburnt core from the burning surface. This charred layer consists of cracked charcoal (Fig. 4.1) with no significant strength or stiffness, while beneath it lies the pyrolysis zone, where active thermal degradation and charring occur. The boundary between these two zones is commonly referred to as the char-line, which is typically defined at the 300 °C isotherm. The distance from the original outer surface of the timber to this char-line is known as the charring depth, and the rate at which this depth progresses over time is termed the charring rate.

The one-dimensional charring rate of wood, as defined in Eurocode 5 Part 1–2, refers to the rate at which the char-line progresses into the wood under fire exposure, assuming heat transfer occurs in a single direction. For the one-dimensional charring rate, calculation of cross-sectional properties should be based on the actual charring depth and the corner roundings. The notional charring rate is an equivalent charring rate that allows the use of an equivalent rectangular residual cross-section.

The main objective of this chapter is to present the experimental work conducted in determining the charring rate of laminated veneer lumber from Malaysian tropical timber exposed to one-dimensional fire in accordance with BS 476: Part 20 (2014) (like EN 1363-1 and ISO 834).

Fig. 4.1 Charred surface after 60 min fire exposure showing the crack charcoal; **a** Dark Red Meranti, **b** Kedondong, **c** Light Red Meranti, **d** Jelutong

Table 4.1 Timber species for LVL

Species	Kasai	Mengkulang	Rubberwood	Eucalyptus	Kedondong
Average density (kg/m^3)	637.4	618	614.9	562.6	448.8

4.2 Testing Material, Specimens, and Equipment

In determining one-dimensional charring rate, several processes needed to be prepared, such as materials, instruments, equipment, and specimens.

4.2.1 Laminated Veneer Lumber

The LVL panels were manufactured in accordance with MS 2209: 2009 (Confirmed: 2020) at one of the factories in Malaysia. The LVLs panels of dimension 1200 × 2400 × 50 mm were produced with 22-ply and having 2.4 mm veneer thickness which were bond together using Phenol Formaldehyde (PF). The species used for LVL in this study are Kasai (*Pometia spp.*), Mengkulang (*Heritiera spp.*), Rubberwood (*Hevea brasiliensis*), Eucalyptus (*Eucalyptus spp.*), and Kedondong (*Canarium spp.*) with densities as shown in Table 4.1.

4.2.2 Furnace Preparation

The furnace used is as shown in Fig. 4.2 that allows the furnace cover to be placed vertically at the front position of the furnace. The furnace is approximately 1500 × 1500 × 1500 mm in dimension fueled by three liquefied petroleum gas (LPG) injection burners with three numbers of built-in thermocouples in the furnace, for monitoring the furnace temperature throughout the fire test.

4.2 Testing Material, Specimens, and Equipment

Fig. 4.2 Furnace; **a** front view, **b** fitted thermocouples to measure the temperature inside the furnace

4.2.3 Furnace Cover Preparation

To securely mount the specimens on the furnace cover to be attached to the furnace, the cover itself was filled with a custom-made castable refractory cement chosen for its ability to withstand high temperatures and provide structural integrity during a fire, complete with three openings with a slot size of 130 mm (W) × 1500 mm (L), as shown in Fig. 4.3. Each furnace cover has three openings designed to accommodate the specimens horizontally (Fig. 4.4).

4.2.4 Preparation of Specimen and Data Acquisition Equipment

The charring rate was determined by analyzing the temperature data obtained from the thermocouples, specifically by tracking the movement of the 300 °C isotherm, which represents the char front. This approach is known as indirect measurement because the charring rate is not directly observed but rather inferred from the temperature profile within the specimen.

Indirect measurement using thermocouples offers several advantages over direct measurement, which involves comparing the original and residual cross-sections. These advantages include continuous monitoring of the charring process and the ability to capture the dynamic behavior of the char front. This method provides accurate, real-time data on the charring rate throughout the fire test, allowing for a detailed analysis of the burning process.

To measure the temperature within the specimens, Type-K thermocouples were inserted through a series of vertically drilled holes at equal intervals, extending from the top surface toward the unexposed side of each specimen. The location of

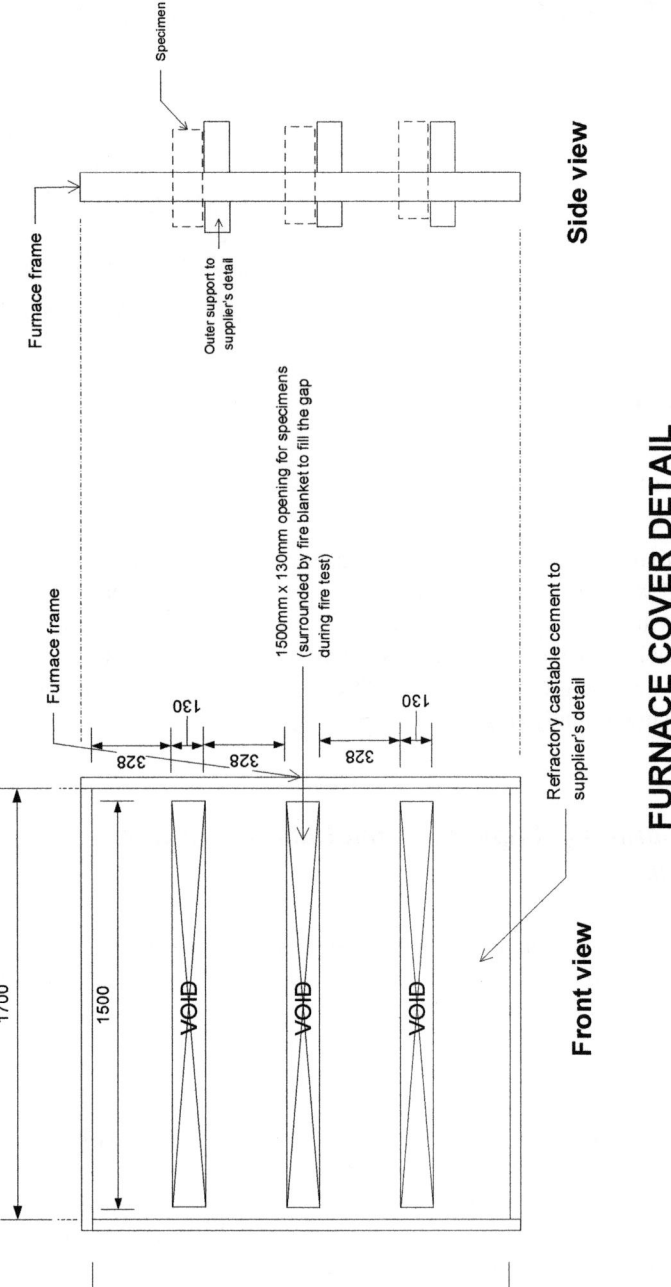

Fig. 4.3 Schematic diagram of furnace cover with three slots for specimen mounting

4.2 Testing Material, Specimens, and Equipment

Fig. 4.4 Actual furnace cover with three slots

thermocouples was 6 mm apart. and placed on the three stations (1, 2, and 3) to obtain comprehensive readings and arranged at different depth of specimen (see Fig. 4.5). All specimens were drilled in the same orientation to ensure consistency. A total of six (6) thermocouples were inserted into the holes, as illustrated in Fig. 4.5.

Once the thermocouples were inserted, the specimens were placed into the slot in the furnace cover, as shown in Fig. 4.7a, which shows the specimens positioned in the slots. To facilitate one-dimensional heat transfer through the LVL members, the LVL test specimens were aligned flush with the cement wall (Fig. 4.7a).

Additionally, any gaps between the specimens and slots were filled with layers of ceramic fiber, acting as an insulating fire blanket. This measure not only prevented fire from escaping through the slots but also minimized smoke generation during the fire test. To securely position the samples onto the frame, heavy-duty metal clamps were used, as shown in Fig. 4.7b. The thermocouples attached to the specimens were then connected to thermocouple wires and compensating cables, which were linked to a multi-channel data logger to continuously record temperature readings throughout the fire test (Fig. 4.8).

During the fire test, temperature measurements were taken at each thermocouple location at 1 min intervals and automatically recorded by the data logger.

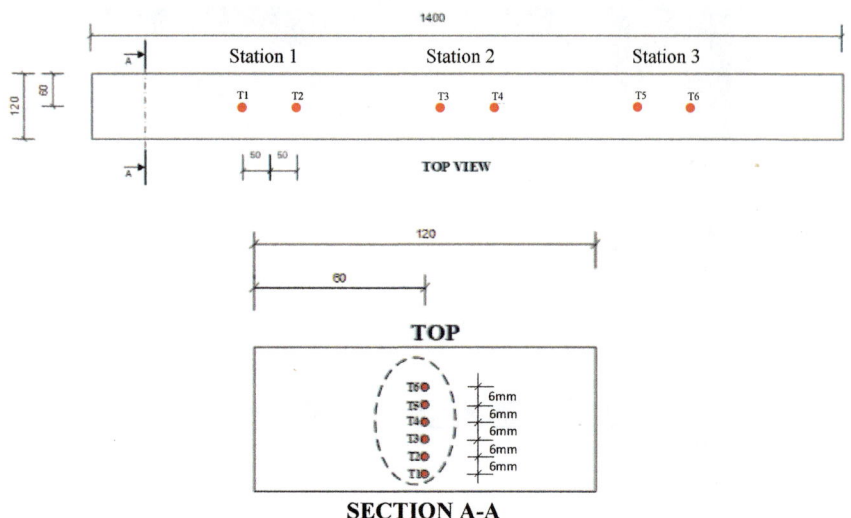

Fig. 4.5 Drilling pattern and location of thermocouples

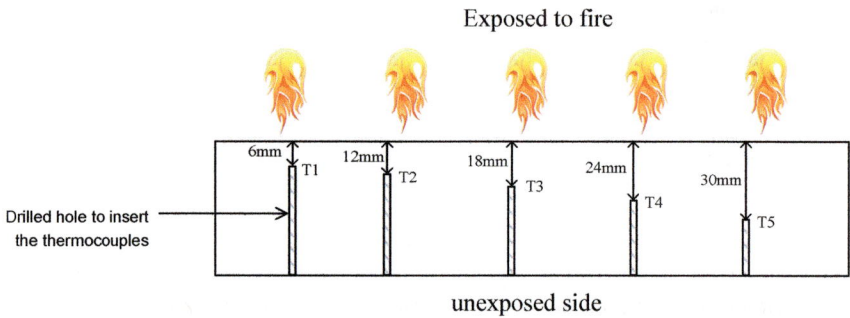

Fig. 4.6 Schematic diagram showing the depth positions of thermocouples

Fig. 4.7 Preparation of testing frame; **a** the gap filled with the fire blanket, **b** securing the specimen with G-clamp

Fig. 4.8 Connecting thermocouples to the electronic data logger

4.3 Visual Observations

The visual observations were made during the one-dimensional fire tests. To shows the typical behavior of LVL under fire exposure, the test on Kedondong LVL is used as an example. The general behavior of the Kedondong LVL during the fire test is summarized in Table 4.2 which presents key observations at different time intervals. In general, all the LVL species exhibited similar behavior under fire exposure, characterized by initial drying, followed by ignition and charring, with the char layer gradually increasing in depth over time.

The fire exposure for the one-dimensional fire test was set to 45 min, consistent with previous research (Aguanno 2013; Peng 2010; Tsai 2010a, b), which indicated that this duration is sufficient to observe significant charring development in LVL without complete combustion. This duration also prevented the specimens from experiencing two-dimensional fire exposure (i.e., exposure on more than one face), which would have occurred if the test had continued beyond 45 min. Limiting the fire exposure to 45 min also ensured that the specimens were not completely consumed, allowing for subsequent measurement of the char depth. It was observed that after 45 min of fire exposure, all specimens were completely charred and continued to burn after removal from the furnace. During the test, some char layers detached, and the specimens deformed due to surface shrinkage at the elevated temperatures.

4.4 Determining the Residual Cross-Section: Measurement and Analysis

At the end of the fire test, all specimens were removed from the furnace and extinguished with water to stop the charring process. Then, the charred material was removed from the specimens, and the residual cross-sections of the specimens were measured. Figure 4.9 shows the conditions of one of the test samples (Kasai species) before and after fire exposure.

The average char depth, d_{char} for each species was calculated by averaging the measurements taken from the three replicate specimens. As a result, the average char

Table 4.2 Visual observations of kedondong LVL during the fire test

Time (mins)	Pictures	Time (mins)	Pictures
0	The test commenced	45	Since the flame was observed on the unexposed side of the specimen, the test was discontinued
3	Flame spreads on the specimen's surface	NA	The char layer fell from the specimens
27	More smoke was released from the gap	NA	Specimens after the fire was extinguished

depths of Rubberwood, Kasai, Mengkulang, Eucalyptus, and Kedondong are 16 mm, 19 mm, 23 mm, 28 mm, and 40 mm, respectively. Figure 4.10 shows enlarged images of the charred cross-sections for each species, highlighting the variations in charring depth.

The non-uniform charring observed in Fig. 4.10 introduces uncertainty into the measurement of the charring rate. The average char depths may not accurately used in the calculation of the charring rate in manual calculation. Therefore, the charring rates derived from direct measurement may have inherent variability and should be interpreted with caution. This highlights the importance of considering indirect measurement using thermocouples, to obtain a more comprehensive understanding of the charring behavior.

4.4 Determining the Residual Cross-Section: Measurement and Analysis

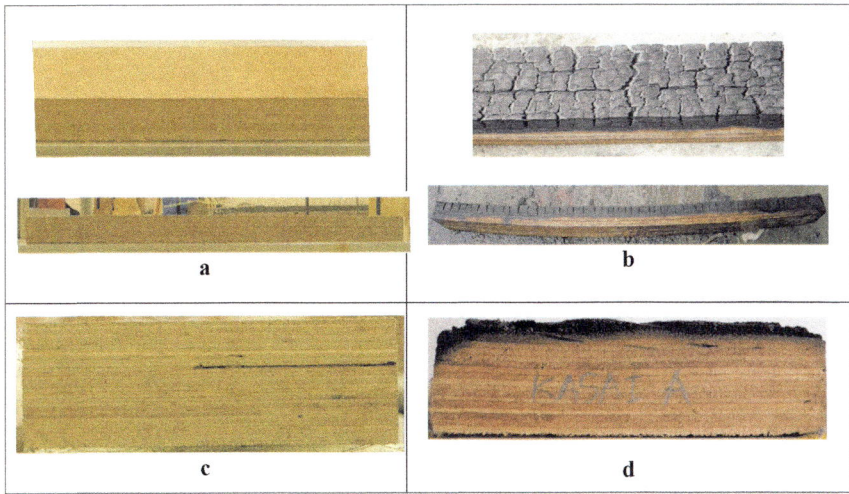

Fig. 4.9 Example of the conditions of the samples before and after fire exposure; **a** side view before fire test, **b** side view after test, **c** front view before test, **d** front view after test

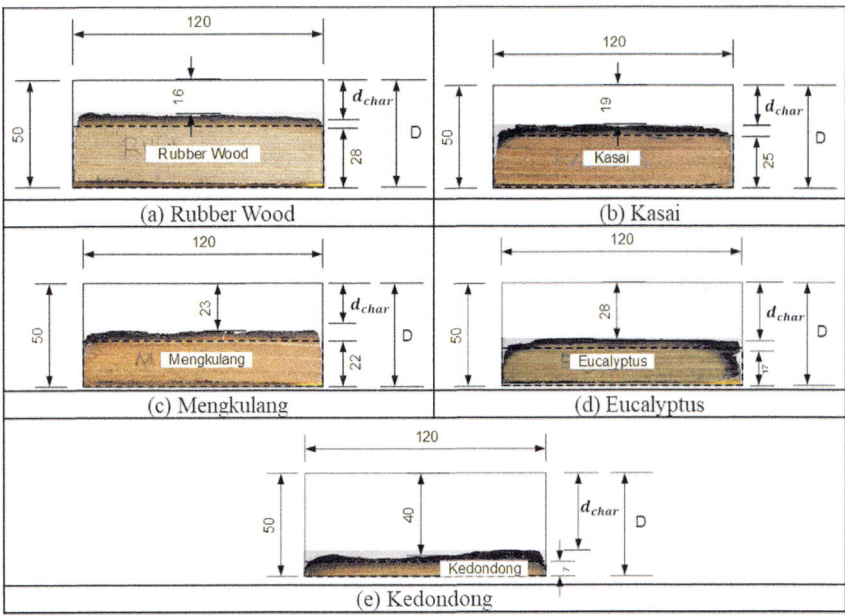

Fig. 4.10 Comparative cross-sectional images of initial and residual specimens analyzed with CAD software Visio 2013, showing average char depths (dimensions in mm)

Table 4.3 Char depth measurement of the residual cross-section

Species	Avg. density (kg/m³)	Avg. char depth, mm
Kasai	637.4	19
Mengkulang	618.0	23
Rubberwood	614.9	16
Eucalyptus	562.6	28
Kedondong	448.8	40

The char depths for each species are presented in Table 4.3. These measurements provide insights into the varying charring behavior of the different species.

4.5 One-Dimensional Charring of LVL Under Standard Fire Exposure

The charring rate of all tested specimens can be obtained indirectly from the temperature data recorded by thermocouples embedded at different depths within the specimens. The 300 °C isotherm, as suggested by Eurocode 5, is assumed to be the boundary between the charred and heated timber. This indirect measurement method provides continuous data on the charring depth throughout the fire test, allowing for accurate tracking of the charring process and analysis of its development.

Based on the guidelines provided in Eurocode 5 (EC5), the one-dimensional charring rate, β_o, is calculated as the char depth, d, divided by the time, t:

$$\beta_o = \frac{d}{t}, \tag{4.1}$$

where
β_o = One-dimensional charring rate.
d = Depth of char.
t = Time.

The one-dimensional charring rate was computed in accordance with BSEN 13,381-7:2019 which outlines a method for determining charring rates from thermocouple data.

4.5.1 Temperature–Time Relationship

The one-dimensional fire tests were stopped when they reached the specified time of 45 min. The thermocouple data were analyzed to plot the 300 °C isotherm, which represents the boundary between the charred and uncharred regions of the LVL, through each specimen for the entire exposure period. Tracking the movement of

4.5 One-Dimensional Charring of LVL Under Standard Fire Exposure 73

this isotherm allows for the determination of the charring rate. Figures 4.11, 4.12, 4.13, 4.14, 4.15 show the temperature–time relationships at different depths within the LVL specimens for all species, with three replicates for each species. These figures illustrate the temperature profiles and how they evolve over time during the fire test.

All temperature–time curves follow a similar trend, characterized by an initial rapid rise in temperature, followed by a gradual leveling off as the char layer

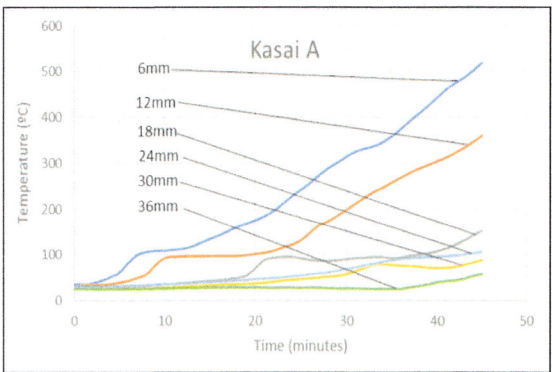

Fig. 4.11 Temperature–time relationships of Kasai

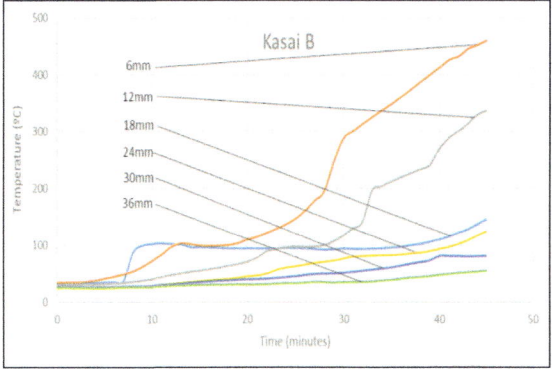

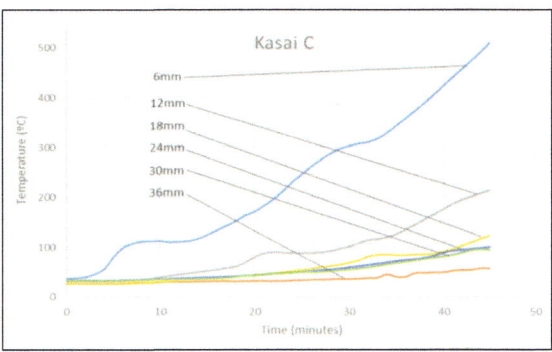

Fig. 4.12 Temperature–time relationships of Mengkulang

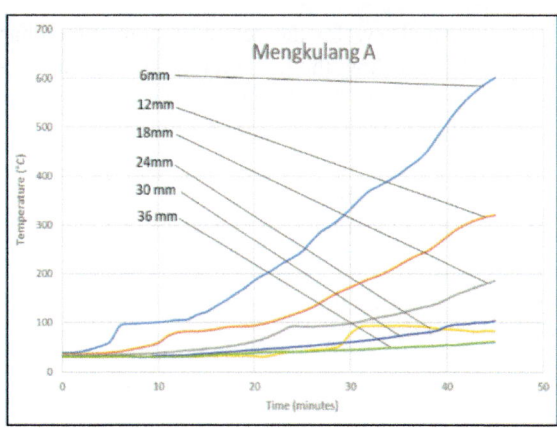

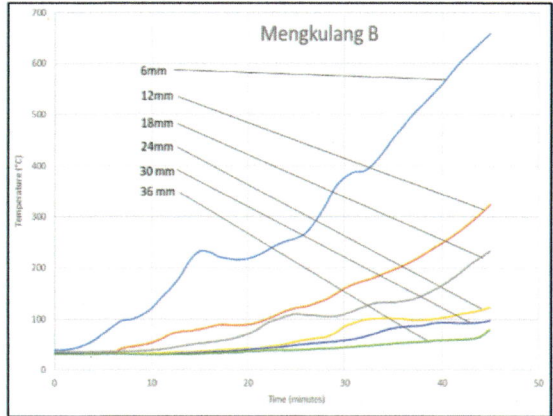

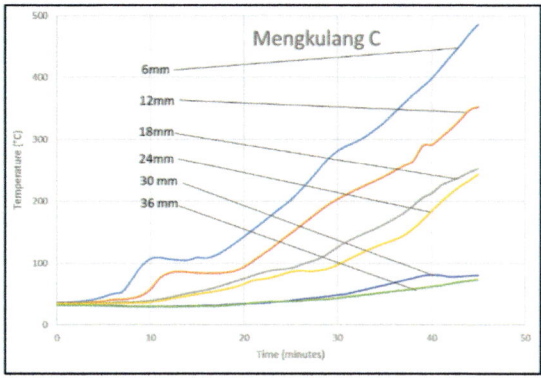

4.5 One-Dimensional Charring of LVL Under Standard Fire Exposure

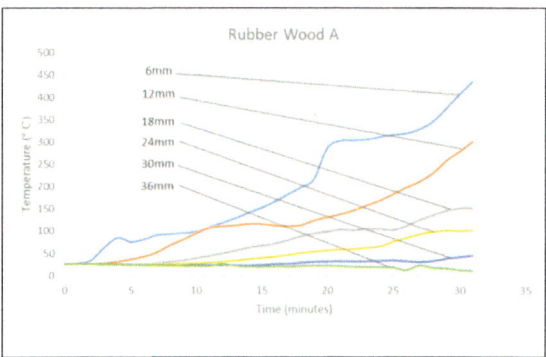

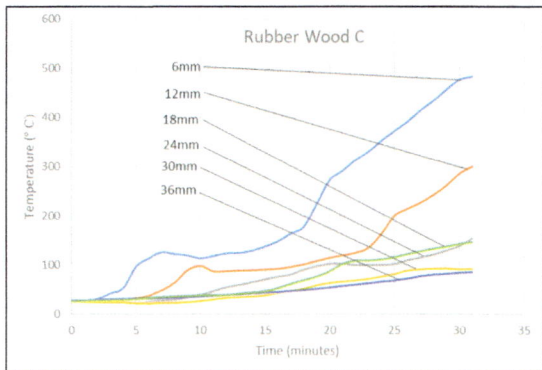

Fig. 4.13 Temperature–time relationships of Rubberwood

develops and provides insulation. Temperature is inversely proportional to char depth, as the char depth increases, the temperature within the specimen decreases. This confirms that thermocouples positioned closer to the fire-exposed surface record higher temperatures. Additionally, as the temperature rises, the shape of the temperature–time curves changes, particularly for the thermocouple closest to the exposed surface. The curves transition from smooth to irregular, with some displaying sudden jumps or fluctuations in temperature. These irregularities may result from variations

Fig. 4.14 Temperature–time relationships of Eucalyptus

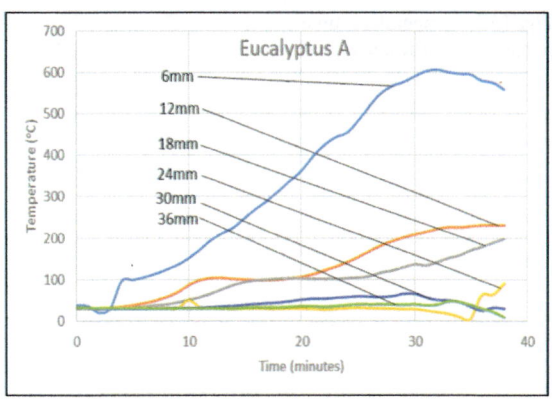

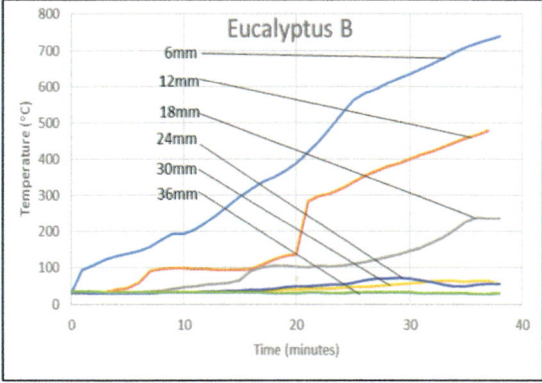

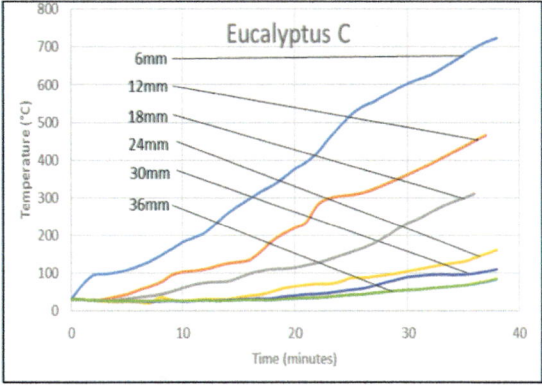

4.5 One-Dimensional Charring of LVL Under Standard Fire Exposure

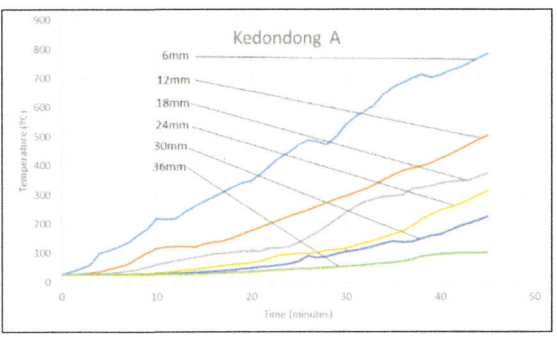

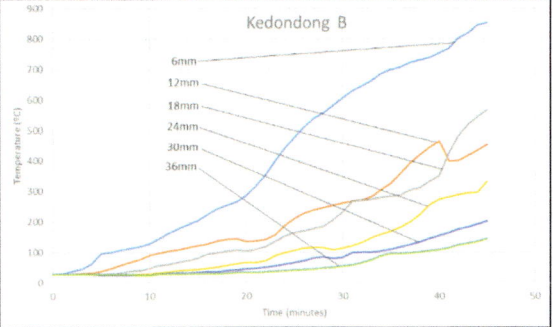

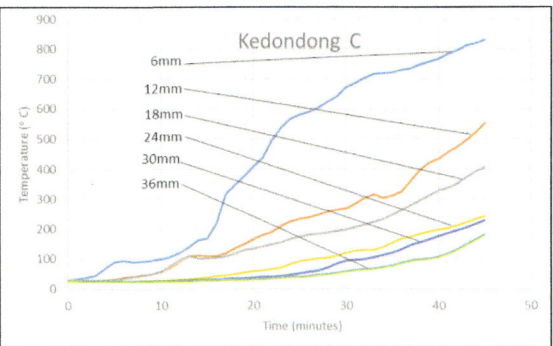

Fig. 4.15 Temperature–time relationships of Kedondong

in the fire's heat release rate, localized cracking or splitting of the char layer, or other factors influencing the heat transfer process.

It can also be observed that the temperature–time curves are not entirely smooth and exhibit sudden increases and fluctuations. Around the 10 min mark, the rate of temperature increase slows down, forming a plateau at approximately 100 °C in the temperature–time curve. During the early phase of the fire, the temperature–time curve for the thermocouple closest to the exposed surface remains relatively flat due to the heat absorption by the laminated veneer lumber (LVL). This initial absorption phase explains why the temperature does not rise rapidly in the early stages of the fire, as the heat Fire is primarily used to evaporate the moisture content within the LVL.

The observed plateau at around 100 °C can be attributed to the moisture present in the LVL. Since 100 °C is the boiling point of water, the heat energy is first used to convert this moisture into steam before significantly increasing the material's temperature. This aligns with the well-known phenomenon of water vaporizing at its boiling point. Figure 4.16 illustrates the evaporation of moisture from the LVL sample during the fire test, as evidenced by the release of steam. The porous nature of the timber allows heat to penetrate easily, leading to a gradual increase in temperature and eventually resulting in ignition and burning of the LVL.

When a timber layer undergoes significant charring, its surface forms a layer of char, which primarily consists of carbon and has lower thermal conductivity than the original wood. This char layer acts as an insulating barrier, hindering the transfer of heat into the inner layers of the LVL and slowing down the charring process. The observed temperature drops in Eucalyptus A after 600 °C could be partly attributed to the insulating effect of the char layer, although other factors might also be involved.

According to Eurocode 5, which provides design guidelines for timber structures, at the 300 °C isotherm, timber is considered no longer able to sustain loads and also has no significant strength and stiffness, meaning that it can no longer effectively support loads or maintain its structural integrity, as discussed by many researchers (Cachim and Franssen 2010; EC5 2004; Thi, Khelifa, El Ganaoui, and Rogaume 2016; Tsai 2010a, b).

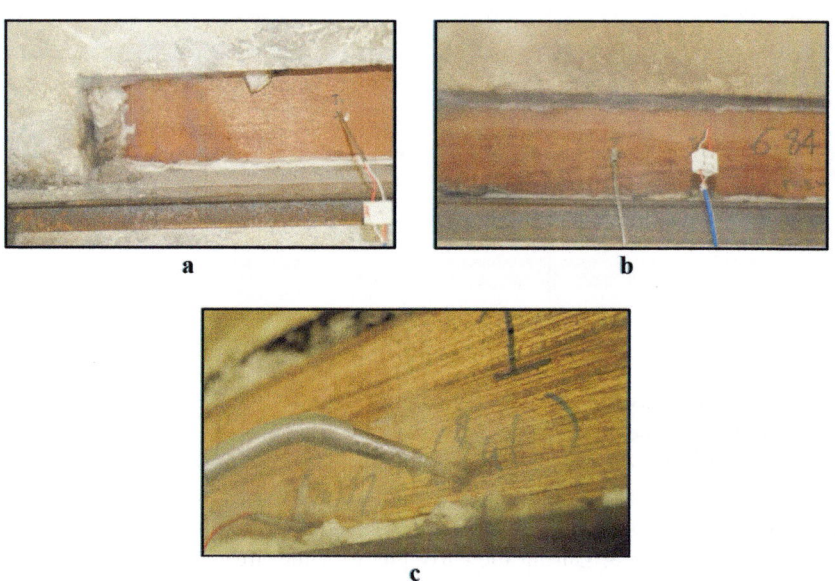

Fig. 4.16 Typical examples of water evaporation in samples during fire exposure; **a** Kasai, **b** Mengkulang, **c** at Thermocouples

4.5.2 Charring Rate Calculation

The determination of the charring rate follows the guidelines set by EN 13,381-7:2019. According to this standard, the charring rate of timber begins at the 300 °C isotherm. However, since the thermocouples are positioned at discrete depths, the exact moment when the temperature reaches 300 °C at a specific location may not be directly recorded.

To address this, interpolation was performed to estimate the precise time at which each thermocouple reached the 300 °C isotherm. Specifically, linear interpolation was used to determine the time ($t_{300}i,j$) at which the temperature reached 300 °C between two consecutive thermocouple readings. Table 4.4 provides an example of the charring rate calculation for the Mengkulang species, based on three replicates. As specified in EN 13,381-7, the charring rate is taken as the maximum value of all charring rates (β_j) to ensure a conservative estimate of charring behavior. Table 4.4 presents the calculated β_j values for Mengkulang at different measurement stations, with the maximum charring rate of 0.46 mm/min observed in specimen C.

In summary, the charring rates for all tested species are provided in Table 4.5. All tested LVL species had densities exceeding 480 kg/m^3, except for Kedondong (~450 kg/m^3). This density threshold is significant because Eurocode 5 recommends specific charring rate values for LVL species with densities above this limit.

Table 4.4 Example calculation of charring rate for Mengkulang

Group	Specimen A-MENGKULANG					
Thermocouple	T1	T2	T3	T4	T5	T6
time [min]/depth [mm]	6	12	18	24	30	36
t_{300} [min]	28.0	41.8	0	0	0	0
$\beta_{i,j}$ [mm/min]	0.21	0.44	0.00	0.00	0.00	0.00
β_j [mm/min]	0.33					
Group	Specimen B-MENGKULANG					
Thermocouple	T1	T2	T3	T4	T5	T6
time [min]/depth [mm]	6	12	18	24	30	36
t_{300} [min]	27.3	47.3	0	0	0	0
$\beta_{i,j}$ [mm/min]	0.22	0.37	0.00	0.00	0.00	0.00
β_j [mm/min]	0.29					
Group	Specimen C-MENGKULANG					
Thermocouple	T1	T2	T3	T4	T5	T6
time [min]/depth [mm]	6	12	18	24	30	36
t_{300} [min]	32.5	40.6	0	0	0	0
$\beta_{i,j}$ [mm/min]	0.18	0.74	0.00	0.00	0.00	0.00
β_j [mm/min]	0.46					

Table 4.5 Average charring rates from experimental results (via thermocouple readings) for specimens with density greater than 480 kg/m^3 (excluding Kedondong)

Species	Avg. density (kg/m^3)	[a]Avg. moisture content (%)	Avg. charring rate (mm/min)
Kasai	637.4	26.4	0.39
Mengkulang	618.0	22.5	0.46
Rubberwood	614.9	21.0	0.48
Eucalyptus	562.6	18.6	0.57
Kedondong	448.8	24.6	0.86

[a] The moisture content taken by moisture meter

Table 4.5 presents the charring rates for different LVL species with varying densities. The table indicates that as density increases, the charring rate decreases, confirming the inverse relationship between density and charring rate. As observed, the charring rate for LVL from Malaysian tropical timber in this study was 0.57 mm/min, which is lower than the 0.65 mm/min charring rate suggested by EC5 for timber with a density greater than 480 kg/m^3.

The charring rate initially increases at each depth after the formation of the initial char layer and then subsequently decreases. This increase in charring rate can be attributed to the rising fire temperature, which accelerates the charring process once the protective char layer is no longer intact. Consequently, the falling off charred layers observed during the fire tests contributed to a temporary increase in the charring rate of the LVL.

4.6 Comparison of Charring Rates with the Eurocode 5

The one-dimensional charring rate values proposed in EC5 have been compared with experimental results, as shown in Fig. 4.17, which plots the charring rate values against density for both the measured data and the EC5 recommendations. It can be observed that for densities below 500 kg/m^3, the EC5 recommendation is unsafe, as it underestimates the charring rate, potentially leading to an overestimation of the fire resistance of timber structures with densities below 500 kg/m^3.

Conversely, for densities above 500 kg/m^3, the EC5 recommendation is conservative, as it overestimates the charring rate. However, Kedondong is an exception, as its charring rate exceeds the EC5 recommendation, making it unsafe for LVL with densities below 500 kg/m^3.

These findings suggest that the EC5 recommendations may require revision to incorporate density-specific charring rate values for LVL, particularly for densities below 500 kg/m^3, to ensure safer and more reliable fire safety designs.

4.7 Conclusion

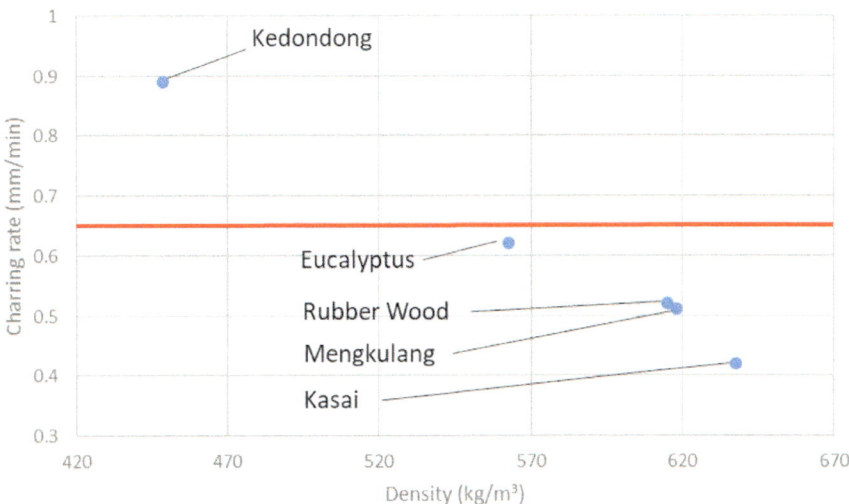

Fig. 4.17 Comparison of experimental results with Eurocode 5

4.7 Conclusion

Based on the observations from these experiments, the following conclusions can be drawn:

a. The EC5 one-dimensional charring rate proposed value (0.65 mm/min) is unsafe for the Kedondong species (densities < 480 kg/m^3) because it underestimates the actual charring rate, potentially leading to an overestimation of the fire resistance of Kedondong LVL. However, for densities above 500 kg/m^3, the EC5 value is slightly conservative, as it overestimates the actual charring rate for those species. This highlights the need for density-specific charring rate values, particularly for lower-density LVL species like Kedondong, to ensure accurate fire safety assessments.

b. The average charring rate for LVL found in this study, based on experimental results (excluding the Kedondong species due to its lower density), is 0.48 mm/min for densities above 480 kg/m^3. This value is lower than the 0.65 mm/min charring rate recommended by EC5. Therefore, a charring rate of 0.48 mm/min can be adopted for LVL made from Malaysian timber species with densities ranging from 500–640 kg/m^3. This lower charring rate suggests that LVL made from Malaysian timber with densities above 500 kg/m^3 may offer better fire resistance than previously estimated by EC5.

References

M. Aguanno, *Fire Resistance Tests on Cross-Laminated Timber Floor Panels: An Experimental and Numerical Analysis*, Master Thesis (Department of Civil and Environmental Engineering Carleton University, 2013)

AS/NZS 1720.4:2019, Timber structures Fire resistance of timber elements, Standards Australia

BS 476: Part 20, BS 476-20:1987 Fire tests on building materials and structures. Method for determination of the fire resistance of elements of construction (general principles) (British Standard, 2014)

P.B. Cachim, J.M. Franssen, Assessment of Eurocode 5 charring XE "charring" rate calculation methods. Fire Technol. **46**(1), 169–181 (2010)

P.C.R. Collier, *Charring Rates of Timber. BRANZ Study Report SR 42, Building Research Association of New Zealand* (1992)

EN 1995-1-2:2004: *Eurocode 5: Design of Timber Structures—Part 1–2: General—Structural Fire Design* (CEN 2004)

EN 363-1:2020, *Fire Resistance Tests—General Requirements* (CEN 2020)

ISO 834-11:2014, Fire resistance tests—Elements of building construction, Part 11: Specific requirements for the assessment of fire protection to structural steel elements

J. König, L. Walleij, One-dimensional charring of timber exposed to standard and parametric fires in initially protected and non-protected protection situations. Tratek Swedish Institute for Wood Technology Research Report No. I9908029 (1999)

MS 544-9-1:2024, Section 1, Code of practice for structural use of timber: Part 9: Fire resistance of timber structures: Section 1: Method of calculating fire resistance of timber members

National Economic Advisory Council, *National Economic Advisory Council New Economic Model for Malaysia* (National Economic Advisory Council, 2010)

L. Peng, *Performance of Heavy Timber Connections in Fire*. PhD Thesis (Carleton University, Canada, 2010)

V.D. Thi, M. Khelifa, M. El Ganaoui, Y. Rogaume, Finite element modelling of the pyrolysis XE "pyrolysis" of wet wood subjected to fire XE "fire." Fire Saf. J. **81**, 85–96 (2016)

W.K. Tsai, *Charring Rates for Different Cross Sections of Laminated Veneer Lumber (LVL)*. PhD Thesis (University of Canterbury, New Zealand, 2010)

R.H. White, E.V. Nordheim, *Charring Rate of Wood for ASTM E 119 Exposure*, (February), 5–30 (1992)

T.-H. Yang, S.-Y. Wang, M.-J. Tsai, C.-Y. Lin, The charring XE "charring" depth and charring rate of glued laminated timber after a standard fire exposure XE "standard fire exposure" Test. Build. Environ. **44**(2), 231–236 (2009)

Chapter 5
Two-Dimensional Charring Rate for Laminated Veneer Lumber from Malaysian Tropical Timber

5.1 Introduction

During a fire, structural elements such as beams are often exposed to complex thermal conditions that can significantly impact their integrity and performance. In many real-world scenarios, beams are subjected to fire from multiple sides, leading to a two-dimensional or notional charring effect. Unlike one-dimensional charring, which occurs when fire exposure is limited to a single surface, two-dimensional charring affects multiple surfaces simultaneously, resulting in a more intricate heat transfer process and structural response.

The charring rate can vary significantly among different wood species, with typical rates ranging from 0.5 to 1.0 mm/min for various timber structures (Fedotov et al. 2022). This variability underscores the importance of conducting comprehensive studies to establish standardized charring rates for different wood types under controlled fire conditions.

This chapter explores the influence of two-dimensional fire exposure on beams, examining how the distribution of fire across multiple surfaces affects the charring rates of laminated veneer lumber (LVL) with different species and densities. The fire tests were conducted in accordance with BS 476: Part 20 (2014) (like EN 1363–1 and ISO 834).

5.2 Testing Material, Specimens, and Equipment

The methodology for testing the two-dimensional fire exposure of wood involves a systematic approach that encompasses sample preparation, fire exposure conditions, measurement techniques, and data analysis. This methodology is crucial for understanding the charring behavior and fire resistance of wood materials under realistic fire scenarios.

5.2.1 Preparation of Specimen

Since the available LVL panels produced by the factory were only 50 mm wide, the 100 mm wide specimens were constructed by gluing two 50 mm panels together using a Phenol-Resorcinol–Formaldehyde (PRF) adhesive, as shown in Fig. 5.1. Table 5.1 provides information on the species, densities, moisture content (measured using a moisture meter), and the number of specimens used in the study.

Following the same procedure as in the one-dimensional test (Chap. 4), a series of 6 mm holes were drilled to various depths vertically using a radial drilling machine, as shown in Fig. 5.2, for thermocouple insertion. Figure 5.3 presents schematic diagrams of the thermocouple positions, including those placed at the corners to specifically measure temperature in these regions. The LVL specimens were exposed to fire on three sides.

The setup for the furnace and testing frame are like one-dimensional fire test. For this two-dimensional fire test, the sample cross-sections were 100 mm width ×

Fig. 5.1 Preparation of 100 mm width LVL; **a** gluing two pieces of 50 mm LVL, **b** the clamping process

5.2 Testing Material, Specimens, and Equipment

Table 5.1 Material physical properties for two-dimensional β_n fire test for 100 mm width specimens

No	Species	Quantity	Density, avg. (kg/m^3)	Moisture content, avg. (%)
1	Rubberwood	3	724.1	21.0
2	Eucalyptus	3	719.2	18.6
3	Kasai	3	653.9	26.4
4	Mengkulang	3	552.8	22.5
5	Kedondong	3	540.6	24.6

Fig. 5.2 Drilling for thermocouple insertion in a two-dimensional fire test of a 100 mm wide specimen

130 mm depth, where the 130 mm deep sections of the specimens were exposed to the fire inside the furnace (Fig. 5.4a). The specimens were connected to the thermocouples as shown in Fig. 5.4b.

The LVL specimens were exposed to fire for 30 min to allow observation of the char layer without being fully burned out. Figure 5.5 shows the specimens before and after 30 min of fire exposure.

5.2.2 Visual Observations

To present the visual observation during and after fire test, the physical performance of LVL from Kasai species was used here as shown in Table 5.2.

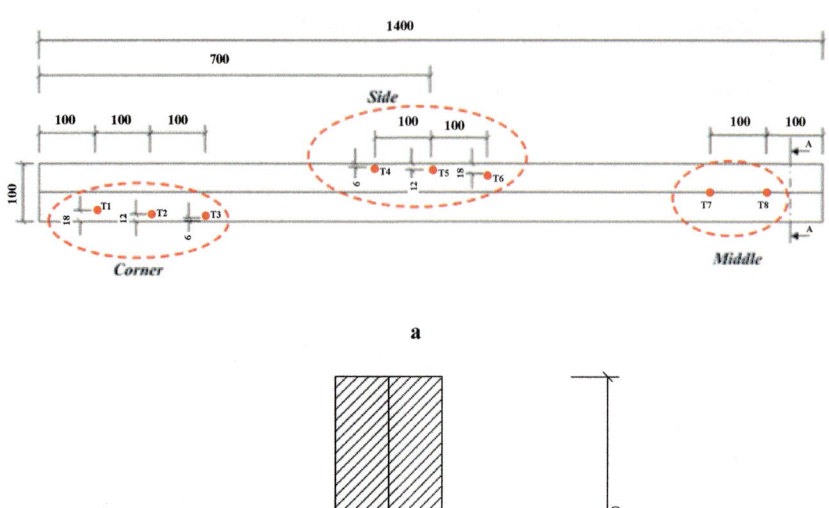

Fig. 5.3 Thermocouple layout and dimensions for a 100 mm width LVL specimen; **a** layout plan, **b** Section A-A

Fig. 5.4 Preparation of two-dimensional fire test; **a** side view showing the 130 mm deep LVL which will be exposed to fire, **b** test frame ready for testing

5.3 Determining the Residual Cross-Section: Measurement and Analysis

Fig. 5.5 Example of the conditions of the samples; **a** before, **b** after fire exposure

5.3 Determining the Residual Cross-Section: Measurement and Analysis

After the fire test, the charred layer of the specimens was physically removed using a scraper. Once the entire specimens were scraped, they were sliced at 20 mm intervals to determine the average char depth, as shown in Fig. 5.6.

The charred layer on the samples was removed to determine the char depth. Figure 5.7 shows the specimens after the removal of char, using Kasai LVL as an example. The width and height of the residual sections were measured on all three sides.

A picture was taken, and the detailed area border lines were retraced using CAD software (Visio 2013). The area for char depth determination was divided into two sections: Sects. 5.1 and 5.2 (Technical Guideline for Europe 2010) (see Fig. 5.8).

Next, to determine the char depth value for the notional charring rate (β_n), the thickness of the charred depth (d_{char}) was measured by averaging the values from both sections ($d_{char,1}$ and $d_{char,2}$), as shown in Fig. 5.8, for each specimen and then compared with the original cross-section. The notional charring rates were introduced to transform the irregular shape of the residual cross-sections into simple rectangular cross-sections, as illustrated in Fig. 5.9. Based on direct measurements of the residual cross-section, the average charring depth over the entire fire duration was then determined.

The sample of char depth calculation as follows:
For Sect. 5.1 (wide side):

$$d_{char,1(avg)} = \frac{t_1 + t_2}{2}. \tag{5.1}$$

For Sect. 5.2 (narrow side):

Table 5.2 Visual observations of Kasai LVL during fire test

Time (mins)	Pictures	Time (mins)	Pictures
0	The test commenced	25	Smoke was continuously released
6	Moisture gradually evaporated	30	The test has been stopped
10	Flame built up	NA	Charred surface from front view
10–20	Char formation was occurring	NA	Side elevation of the specimen after removal from the furnace

5.3 Determining the Residual Cross-Section: Measurement and Analysis

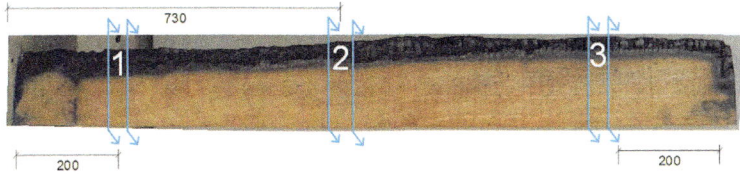

Fig. 5.6 Schematic diagram of the tested samples cut into a series of blocks

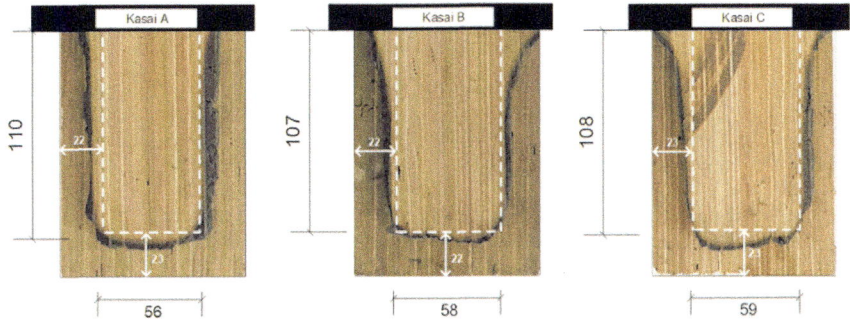

Fig. 5.7 Residual of charred samples for Kasai LVL analyzed with Visio Professional 2013 (dimension in mm)

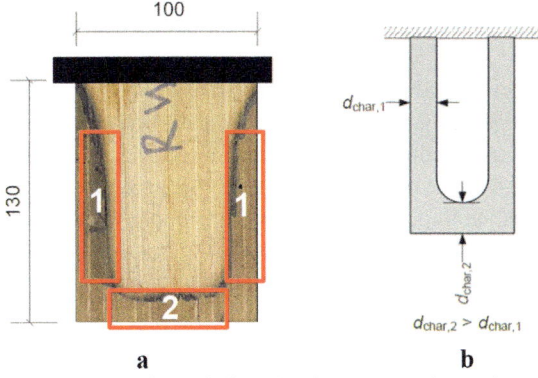

Fig. 5.8 Area for determination of char depths; **a** actual specimens, **b** method of measurement of char depth for three-sided fire exposure

$$d_{char,2} = t_3. \quad (5.2)$$

Subsequently, notional charring rate for each species were calculated based on Eq. (5.3).

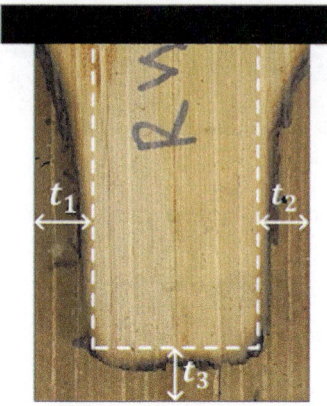

Fig. 5.9 Determination of char depth by averaging from several points analyzed with a CAD software

$$\beta_n = \frac{d}{t}. \qquad (5.3)$$

5.4 Evaluation of Charring Rate Based on Thermocouple Readings

The fire tests for the 100 mm width LVL specimens were stopped when they reached the specified duration of 30 min. The thermocouple data from the 100 mm width LVL tests were analyzed to plot the 300 °C isotherm through each specimen for the entire exposure period. The time at which each thermocouple reached 300 °C was plotted against the corresponding depth of the thermocouple within the specimen. This was done separately for each specimen to track the progression of the 300 °C isotherm. Figure 5.10 shows an example of the temperature–time profile at different depths from exposed surface.

The corner side of the specimen exhibits higher temperature values due to the two-dimensional heat transfer; whereas, the temperature within the central part of LVL increased slowly because the char layer formed on the surface acts as an insulator, slowing down the heat penetration into the inner layers of the LVL. In this study, the mean temperature at the inner LVL ranged from 28.9 to 60 °C (Tc7-Tc8 measured points). This indicates that the temperature at the center of all tested LVL specimens remained relatively stable, with minimal influence from the fire exposure. Fire tests with LVL beams exposed to fire on three sides clearly showed that the charring rate increases rapidly as the cross-section width decreases below 100 mm. The temperature–time line for 100 mm width LVLs also exhibit a similar pattern, with a plateau at 100 °C which due to moisture evaporation.

5.5 Comparison of Charring Rates for LVL with the Eurocode 5

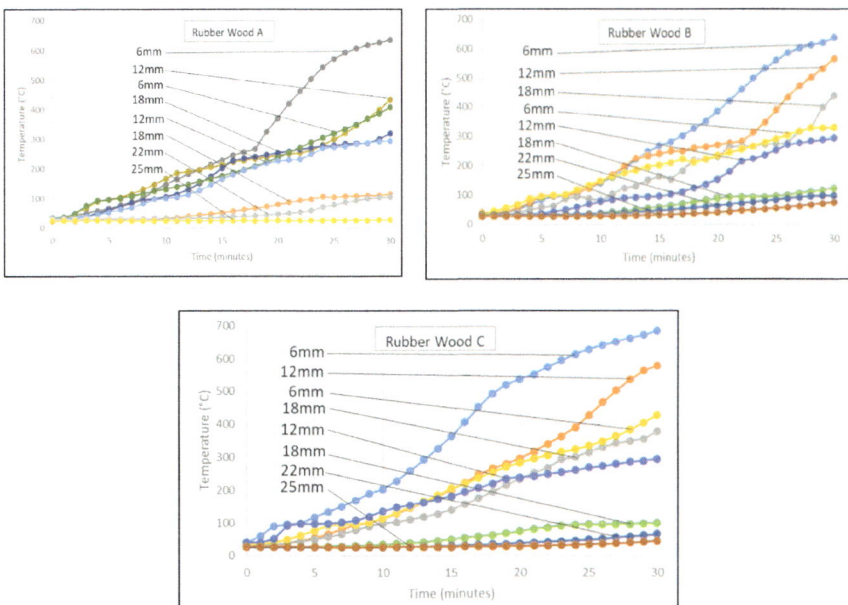

Fig. 5.10 Temperature–time profile for Rubberwood LVL at different depths from exposed surface

Table 5.3 shows the depths at which the thermocouples reached the 300 °C isotherm for all LVL specimens. Analyzing the 300 °C isotherm allows for the determination of the charring rate, which is a crucial parameter for predicting the fire resistance of timber structures.

The charring rate was calculated in accordance with prEN 13,381–7:2024 and the examples of the calculation of charring rate are shown in Table 5.4. The results of the charring rate for all species are tabulated in Table 5.5.

Table 5.5 presents the charring rates for different LVL species with varying densities. The data in Table 5.5 clearly shows the effect of density on charring rate performance. The charring rate values are smaller for higher density species and increase as the density decreases. After a 30 min fire exposure, the charring depths of LVL specimens decreased in the following order: Rubberwood > Eucalyptus > Kasai > Mengkulang > Kedondong.

5.5 Comparison of Charring Rates for LVL with the Eurocode 5

Figure 5.11 illustrates the relationship between the charring rate of various LVL species and the charring rate specified for LVL in Eurocode 5 (EC5). According to the figure, Rubberwood and Eucalyptus (with a density greater than 700 kg/m^3) fall

Table 5.3 Thermocouples reached the 300 °C isotherm at different depths for all species in the 100 mm width LVL specimens

Species	Corner			Wide Side			Middle	
	Tc1	Tc2	Tc3	Tc4	Tc5	Tc6	Tc7	Tc8
	6 mm	12 mm	18 mm	6 mm	12 mm	18 mm	22 mm	25 mm
Kasai A	✓	✓	✓	✓	✓	✗	✗	✗
Kasai B	✓	✓	✓	✓	✗	✗	✗	✗
Kasai C	✓	✓	✓	✓	✗	✗	✗	✗
Mengkulang A	✓	✓	✓	✓	✓	✗	✗	✗
Mengkulang B	✓	✓	✓	✓	✓	✗	✗	✗
Mengkulang C	✓	✓	✓	✓	✓	✗	✗	✗
Rubberwood A	✓	✓	✓	✓	✓	✗	✗	✗
Rubberwood B	✓	✓	✓	✓	✓	✗	✗	✗
Rubberwood C	✓	✓	✓	✓	✓	✗	✗	✗
Eucalyptus A	✓	✓	✓	✓	✓	✗	✗	✗
Eucalyptus B	✓	✓	✓	✓	✓	✗	✗	✗
Eucalyptus C	✓	✓	✓	✓	✓	✗	✗	✗
Kedondong A	✓	✓	✓	✓	✓	✗	✗	✗
Kedondong B	✓	✓	✓	✓	✓	✗	✗	✗
Kedondong C	✓	✓	✓	✓	✓	✗	✗	✗

✓ Achieved 300 °C.
✗ Did not achieved 300 °C
Tc Thermocouple

within the EC5 recommendation, which sets a maximum charring rate of 0.7 mm/min for LVL with a density exceeding 480 kg/m^3.

However, despite having densities above 480 kg/m^3, Kedondong, Mengkulang, and Kasai LVLs exhibit charring rates higher than 0.7 mm/min. This suggests that the charring behavior of LVL produced from tropical hardwood species may not fully align with the existing EC5 limits for safe design. Therefore, further investigation is necessary to establish more accurate charring rate guidelines for these materials.

5.6 Conclusions

Based on observations from these experiments, the following conclusions can be drawn for LVL from tropical hardwoods:

a. The average charring depth and charring rate were highest at the corners, compared to the middle and wider sides of the LVL.
b. A clear inverse relationship was observed between the average charring depth and density, indicating that denser LVL exhibited lower charring depths.

5.6 Conclusions

Table 5.4 Calculation of charring rate by indirect measurement for 100 mm width LVL

Group	Specimen A-Rubberwood							
	Corner			Middle		Side		
Thermocouple	Tc1	Tc2	Tc3	Tc7	Tc8	Tc4	Tc5	Tc6
time [min]/depth [mm]	6	12	18	22	25	6	12	18
t_{300} [min]	18.6	24.9	28.9	0	0	23.7	0	0
$\beta_{i,j}$ [mm/min]	0.32	0.95	1.50	0.00	0.00	0.25	0.00	0.00
β_j [mm/min]	0.92			0.00		0.25		
$\beta_{j(total)}$ [mm/min]	0.58							
Group	Specimen B-Rubberwood							
	Corner			Middle		Side		
Thermocouple	Tc1	Tc2	Tc3	Tc7	Tc8	Tc4	Tc5	Tc6
Time [min]/depth [mm]	6	12	18	22	25	6	12	18
t_{300} [min]	16.9	22.4	26.5	0	0	25.4	0	0
$\beta_{i,j}$ [mm/min]	0.36	1.09	1.46	0.00	0.00	0.24	0.00	0.00
β_j [mm/min]	0.97			0.00		0.24		
$\beta_{j(total)}$ [mm/min]	0.60							
Group	Specimen C-Rubberwood							
	Corner			Middle		Side		
Thermocouple	Tc1	Tc2	Tc3	Tc7	Tc8	Tc4	Tc5	Tc6
Time [min]/depth [mm]	6	12	18	22	25	6	12	18
t_{300} [min]	13.2	20.1	24.0	0	0	21.3	30.2	0
$\beta_{i,j}$ [mm/min]	0.45	0.60	1.54	0.00	0.00	0.28	0.40	0.00
β_j [mm/min]	0.86			0.00		0.34		
$\beta_{j(total)}$ [mm/min]	0.60							

Table 5.5 Average charring rates from experimental results for LVL

Species	Avg. density (kg/m^3)	[a]Avg. moisture content (%)	Avg. charring rate (mm/min)
Rubberwood	724.1	21.0	0.60
Eucalyptus	719.2	18.6	0.67
Kasai	653.9	26.4	0.74
Mengkulang	552.8	22.5	0.80
Kedondong	540.6	24.6	0.91

[a] The moisture content was measured using a moisture meter

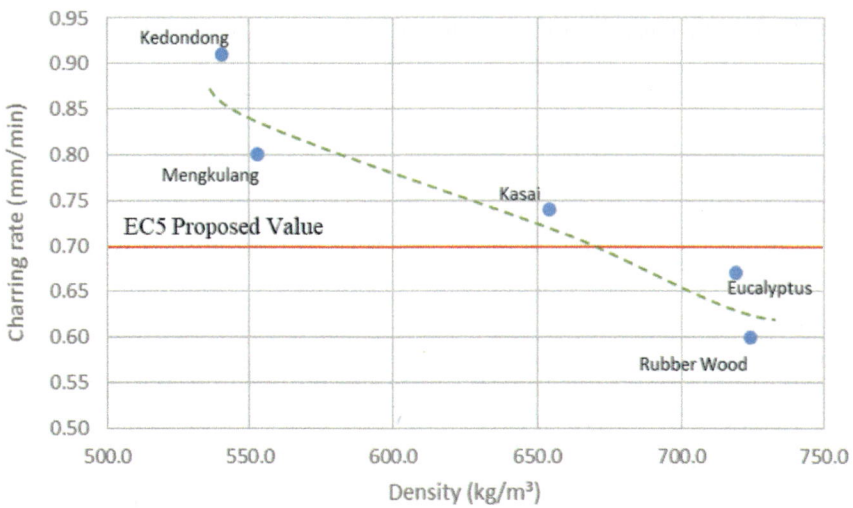

Fig. 5.11 Comparison of experimental results with Eurocode 5

c. The charring rate of LVL made from Malaysian tropical hardwood species followed a decreasing trend in the order of: Rubberwood > Eucalyptus > Kasai > Mengkulang > Kedondong.
d. The charring rate of LVL from Malaysian tropical timber was generally higher than the rate specified in Eurocode 5 (EC5) for LVL with a density above 480 kg/m^3. Only LVL with a density exceeding 700 kg/m^3 fell within the EC5 recommendations.

These findings highlight the need for further investigation into the charring behavior of LVL from tropical hardwood species to ensure accurate design considerations.

References

BS 476: Part 20 (2014) BS 476–20:1987 Fire tests on building materials and structures. Method for determination of the fire resistance of elements of construction (general principles), British Standard
EN 1363–1:2020: Fire resistance tests - Part 1: General requirements, CEN (2020)
EN 1995–1–2:2004: Eurocode 5: Design of timber structures—Part 1–2: General -Structural fire design, CEN (2004)
I.O. Fedotov, A.B. Sivenkov, G. Khasanova, The efficiency of various fire XE "fire" protectants for wooden structures. Eurasian Chem. Technol. J. **24**(1), 33 (2022)
ISO 834–11:2014, Fire resistance tests—Elements of building construction, Part 11: Specific requirements for the assessment of fire protection to structural steel elements. International Standard Organization
prEN 13381–7:2024: Test methods for determining the contribution to the fire resistance of structural members—Part 7: Applied protection to timber members, CEN (2024)

Index

A
Area for char depth, 87
Average char depth, 69, 87

B
Bending strength, 9, 12
Bottom layers, 12
Burn, 26, 33, 69
Burning, 22, 23, 26–30, 32–34, 39, 45, 46, 49, 51, 52, 63, 65, 78

C
Carbonaceous char, 26, 30, 47
Carbon footprint, 2, 14
Carbon monoxide, 29, 30, 32–34
Char, 23, 24, 26, 28, 30, 39, 45, 47, 52–54, 63, 65, 69–73, 78, 80, 85, 87, 89, 90
Char layer, 26, 30, 39, 45, 47, 52, 69, 70, 73, 78, 80, 85, 90
Charring, 39, 46, 47, 52–58, 60, 63–65, 69, 70, 72, 73, 78–81, 83, 87, 89–94
Charring depth, 54–57, 63, 70, 72, 87, 92
Conductivity, 78
Corner rounding, 56

D
Density, 12, 13, 32, 33, 49, 50, 52, 53, 58, 60, 79–81, 91–94
Direct measurement, 65, 70

E
Elevated temperatures, 31, 39, 47, 69

Energy, 21, 23, 27–30, 32–34, 47, 78
Exothermic reaction, 31
Extinguished, 27, 69, 70

F
Fire, 16, 21–23, 25–29, 31, 33, 34, 39–58, 60, 63–65, 67–72, 75, 77, 78, 80, 81, 83–91
Fire exposure, 23, 44, 45, 52–57, 63, 64, 69, 71, 78, 83, 85, 87, 89–91
Fire resistance, 16, 44, 58
Fire test, 48, 50, 51, 57, 64, 65, 67, 69–73, 78, 84–88
Furnace cover, 64–67

G
Grain, 9, 13, 14, 23, 47, 53

H
Hardwood, 60
Heartwood, 24
Heat, 21, 22, 25–34, 39, 40, 42, 43, 45–47, 49, 51, 52, 54–56, 63, 67, 77, 78, 83, 90
Heat output, 31–34
Heat release rate, 21, 27, 43, 77
Heat transfer, 26, 33, 39, 46, 47, 52, 54–56, 63, 67, 77, 83, 90
Hemicellulose, 23–26, 30, 34
Higher density, 91

I
Ignition time, 21
Impermeability, 24
Indirect measurement, 65, 70, 72, 93
Initial absorption, 77
Initial moisture content, 29

J
Joists, 12, 16, 58

K
Kilns, 33

L
Lignin, 23–26, 30, 34
Load-bearing, 11, 12, 14, 15, 42, 54
Lower thermal resistance, 26

M
Malaysian tropical timber, 48, 49, 58, 63, 80, 94
Mass timber, 2
Moisture content, 7, 14, 21, 23, 28, 29, 33, 34, 43, 53, 78, 80, 84, 93

N
Notional charring depth, 56
Notional charring rates, 56, 87

O
One-Dimensional, 52, 54, 72
Orientation, 9, 23, 25, 53, 67
Oxidation, 22, 31, 47

P
Perpendicular to the grain, 23
Porous, 78
Protective, 26, 39, 40, 47, 60, 80
Pyrolysis, 22, 23, 26, 28, 30–34, 45, 47, 52, 53, 63

Pyrolysis zone, 30, 53, 63

Q
Quality of veneer, 10
Quantifies the energy, 27

R
Reaction to fire, 21, 40, 42, 48, 49, 51

S
Smoke production, 21, 22, 33, 44, 45, 48, 49, 51
Smoke release, 50
Softwood, 33, 47, 58
Solid timber, 60
Standard fire exposure, 72
Stiffness, 2, 9, 11, 17, 25, 39, 47, 63, 78

T
Temperature, 9, 14, 23, 26–30, 32–34, 39, 43–47, 53, 54, 64, 65, 67, 72, 73, 77–80, 84, 90
Temperature-time, 72, 75–77, 91
Thermal degradation, 23, 25, 26, 30, 45, 46, 52, 63
Thermocouples, 64, 65, 67–70, 72, 75, 79, 85, 91
Timber cross-section, 54
Tropical hardwood, 4, 92, 94
Two-dimensional, 55, 58, 69, 83–86, 90

V
Vaporization, 45
Volatile, 26, 28, 30–34, 47

W
Weight loss, 30
Weight profile, 33
Wide side, 87